A Student's Guide to Newton's Laws of Motion

Newton's laws of motion, which introduce force and describe how it affects motion, are the gateway to physics – yet they are often misunderstood due to their many subtleties. Based on the author's twenty years of teaching physics and engineering, this intuitive guide to Newton's laws of motion corrects the many misconceptions surrounding this fundamental topic. Adopting an informal and pedagogical approach and a clear, accessible style, this concise text presents Newton's laws in a coherent story of force and motion. Carefully scaffolded everyday examples and full explanations of concepts and equations ensure that all students studying physics develop a deep understanding of Newton's laws of motion.

SANJOY MAHAJAN is Research Affiliate in the Mathematics Department and J-WEL Affiliate at the Jameel World Education Lab at the Massachusetts Institute of Technology. After having studied mathematics at the University of Oxford and physics at the California Institute of Technology, he has taught physics, mathematics, and engineering around the world, including at the African Institute for Mathematical Sciences and the University of Cambridge. He is the author of *The Art of Insight in Science and Engineering* (MIT Press) and *Street-Fighting Mathematics* (MIT Press).

Other books in the Student Guide series:

A Student's Guide to Newton's Laws of Motion

SANJOY MAHAJAN
Massachusetts Institute of Technology

CAMBRIDGE
UNIVERSITY PRESS

CAMBRIDGE
UNIVERSITY PRESS

University Printing House, Cambridge CB2 8BS, United Kingdom

One Liberty Plaza, 20th Floor, New York, NY 10006, USA

477 Williamstown Road, Port Melbourne, VIC 3207, Australia

314-321, 3rd Floor, Plot 3, Splendor Forum, Jasola District Centre, New Delhi - 110025, India

79 Anson Road, #06-04/06, Singapore 079906

Cambridge University Press is part of the University of Cambridge.

It furthers the University's mission by disseminating knowledge in the pursuit of
education, learning and research at the highest international levels of excellence.

www.cambridge.org
Information on this title: www.cambridge.org/9781108471145
DOI: 10.1017/9781108557702

First published 2020

A catalogue record for this publication is available from the British Library

ISBN 978-1-108-47114-5 Hardback
ISBN 978-1-108-45719-4 Paperback

To John William Warren (1923–2016),
Senior Lecturer in Physics and Reader in Physics Education
at Brunel University, London,
whose works set me on
the path to understanding
Newton's enchanting laws of motion

Contents

Preface

Newton's three laws of motion, the basis of almost all science and engineering, are one of the great achievements of human culture. Using them, we explain, predict, and plan the motion of bodies in the natural and in our human-created worlds. Doing so requires knowing how forces affect motion – knowledge embodied in Newton's second law. Your fluency with this law is the ultimate goal of this book. But first you must know when this law is valid – knowledge provided by Newton's first law. And understanding the first law requires a prior idea, interaction – embodied in Newton's third law.

Thus, you will meet the three laws in the following order: (1) the third law, to introduce interaction; (2) the first law, to describe when the second law can even be used; (3) and finally the second law, to describe what forces do.

But, wait! Before studying the effect of force (the second law) or the idea of interaction (the third law), don't you need to know what force is? No one has answered that question fully. Fortunately, we can understand and use Newton's laws without a solution to that philosophical conundrum. All that we need to know is that a force is a push or a pull. Thus, a force has a strength, formally known as its magnitude, and a direction. Mathematically, force is a vector.

Now you are ready for Newton's laws. To help you learn them, I have embedded throughout this book three types of questions. Questions preceded by a rightward-pointing triangle (▶) are from me to you. They are what I would ask you in a one-to-one tutorial on Newton's laws. Questions preceded by a leftward-pointing triangle (◀) are from you to me. They are questions that students have asked or should ask me. For both types of triangle questions, but especially for the questions from me to you (▶), try to answer the question before reading on for my explanation. In that way, you will learn Newton's laws more quickly. (When my explanation is lengthy and the answer itself easy to miss, I point out the answer to the triangle question explicitly.)

The third type of question is end-of-chapter problems. Like traditional home-work problems, they ask you to apply the ideas that you have learned so far (including in earlier chapters!). Their solutions are available online. As with the triangle questions, try your hand before studying my solution, but do use my solutions as worked examples – one of the most effective ways to learn [20].

Newton's laws are subtle. I have been studying them for over 30 years and teaching them for over 20 years. Only now do I understand many of their sub-tleties. This book will help you learn in weeks what I learned over decades, an attempt to fulfill the purpose of teaching described by the physicist Edwin Jaynes: to implant a way of thinking so that you, the student, can "learn in one year what the teacher learned in two" [10]. And you can. For I took detours, covered up deep misunderstandings with symbol manipulation and formalism, and fell into many conceptual traps – traps arising partly from what the physics educator J. W. Warren describes as the "incredible confusion of approach" [25, p. 45]. In the following chapters, we journey quickly and directly to the heart of this fascinating subject.

On y va!

Acknowledgments

I am grateful for help from many sides. The book has been typeset using ConTEXt, built on TEX; the friendly ConTEXt community, including Wolfgang Schuster, Mikael Sundqvist, and Hans Hagen, have offered valuable advice throughout. The Asymptote developers have provided a powerful and enjoyable tool for making scientific figures. At Olin College, the Faculty Development Program provided a writing grant, and Vincent Manno arranged a developmental leave at the right time. Deborah Beers-Jones has taught me about teaching through teaching me piano. Dave Pritchard has for many years shared his wisdom about teaching Newton's laws. Steve Holt and Dan Fleisch made insightful comments on the entire text – as did Joshua Roth, who improved every page. Simon Capelin, Nick Gibbons, and Roisin Munnelly at Cambridge University Press provided valuable guidance throughout, and John King expertly edited the final manuscript. Students in my Mechanics courses helped me clarify many confusing parts. The Art of Insights group at MIT – Sheryl Barnes, Dave Darmofal, Denny Freeman, Woody Flowers, Warren Hoburg, Sanjay Sarma, and Gerry Sussman – offered a stimulating forum to rethink the teaching of physics and engineering. Arthur Eisenkraft introduced me to the fascination of physics. J. W. Warren, in *Understanding Force* [25] and other works, set me on the path to understanding Newton's laws. And Juliet, last in this list but first in my life, encouraged me to become a writer.

1

Newton's Third Law: Forces Belong to Interactions

The most misunderstood and yet the most important of Newton's three laws is the third law. For it introduces the idea of interaction. Without that idea, Newton's first law (Chapter 3), which requires removing interactions, makes no sense. And without the first law, you don't know when you can use Newton's second law (Chapter 4) – the heart of mechanics. For want of an interaction, the kingdom of physics is lost!

Like me, you may have heard or learned the third law in the action–reaction form: "For every action, there is an equal and opposite reaction." That form confused me for 20 years and might do the same to you. Here is a clearer form inspired by the work of the physics educator Cornellis Hellingman [7].

> **Newton's Third Law.** A force is one side of an interaction between two bodies A and B. The interaction acts equally strongly in the two opposite directions, from A to B and from B to A.

This interaction form reminds us, whenever we encounter a force, to look for the interaction to which it belongs and therefore for the two interacting bodies.

We can express the same law in mathematical notation. If the two bodies are A and B, then one force is $\mathbf{F}_{A \text{ on } B}$: the force on body B due to its interaction with body A – or, more concisely, the force of A on B. The other force is $\mathbf{F}_{B \text{ on } A}$: the force on body A due to its interaction with body B – the force of B on A. The boldface type for $\mathbf{F}$ indicates that $\mathbf{F}$ is a vector, meaning that it has magnitude and direction. (In handwriting, where boldface is hard to make, you'll see an italic letter with an arrow: $\vec{F}$.) Newton's third law then says that

$$\mathbf{F}_{A \text{ on } B} = -\mathbf{F}_{B \text{ on } A}. \tag{1.1}$$

The bare minus sign, meaning multiplication by -1, ensures that the two forces have opposite directions and equal magnitudes. The magnitude, in the internationally standard metric system (the SI system), is measured in newtons. In one of history's jokes, this unit was never used by Newton. In Section 1.4, you develop a feel for this unit as we estimate the magnitudes of diverse forces. But first you learn how to use the third law (Section 1.1), learn how to classify forces (Section 1.2), and meet the forces most important in the world around us (Section 1.3). In the final section (Section 1.5), you learn why you should avoid two familiar forces whose use almost inevitably generates confusion.

1.1 Using the Third Law

Figure 1.1 Standing on the ground. You stand on the ground and are pulled downward by the gravitational force $\mathbf{F}_g$. What's $\mathbf{F}_g$'s third-law counterpart force?

Here is an everyday example of Newton's third law and of how it reveals a hidden and surprising aspect of the world. The situation: You stand on the ground, and the earth pulls you with a gravitational force $\mathbf{F}_g$ (Figure 1.1). Although the pull acts on each particle within you, this distributed set of forces is, for most purposes, equivalent to a single force acting at your center of mass – at the dot in the diagram. (Problem 5.8 explores a case where this equivalence isn't valid.)

Now try to answer the following question marked with a triangle. As I mention in the Preface (on p. ix), the rightward-pointing triangle indicates a question that I'd ask you in a one-to-one tutorial on Newton's laws. Even in this less personal written format, do make a decent attempt to answer the question; then compare your answer with the full explanation that follows.

▶ *What's the third-law counterpart force of the gravitational force?*

Most students and many teachers answer this question incorrectly (see, for example, the research by Hellingman [7] and by Terry and Jones [24]). As a student and for many years as a teacher, I would have been among them and would have answered, "the upward force of the ground on me." For I would have used the action–reaction form of Newton's third law and reasoned as follows.

Gravity is trying to pull you into the ground, so the gravitational force pulling you down into the ground must be the action. The ground complains: "Wait! Bowed low by the weight of the world though I may be, I am still a solid object. You shall not pass through me!" In self-defense, it reacts by pushing you upward. This upward force must be the reaction. Thus, it's exactly as strong as the gravitational force. (This conclusion is almost unavoidable in the Commonwealth countries, where any upward force from the ground is called a "reaction force.")

Although the conclusion is right, the reasoning is wrong – the worst combination of right and wrong because the rightness of the conclusion obscures the fundamental error in the reasoning. In our ends-justify-the-means age, the rightness may seem like sufficient justification. However, the same reasoning in many other situations easily leads to wrong conclusions about the upward force from the ground – for example, when someone pushes you downward, when you land after a jump, or when you stand in an accelerating elevator.

Fortunately, you cannot fall into these traps when you use the interaction form of Newton's third law. It's embodied in the following procedure.

1. *Determine what two bodies interact to produce the given force.* Here, the given force is the gravitational force on you. Therefore, the two bodies that interact are you and the earth.

2. *Classify the interaction.* The choice, as you soon learn in Section 1.2.1, is between a gravitational and an electromagnetic interaction. This interaction is gravitational.

3. *Describe the given force as one side of this interaction.* In words, it's the gravitational force of the earth on you. In symbols, it's $\mathbf{F}_{\text{earth on you}}$.

4. *Describe its third-law counterpart force as the other side of the interaction.* You simply reverse the two bodies' roles. Here, "the gravitational force of *the earth* on *you*" becomes "the gravitational force of *you* on *the earth*": $\mathbf{F}_{\text{you on earth}}$.

5. *Remind yourself of how strong the counterpart force is and in what direction it points.* The two forces that constitute the gravitational (or any) interaction are equal and opposite ($\mathbf{F}_{\text{A on B}} = -\mathbf{F}_{\text{B on A}}$), so your gravitational force on the earth has the same strength as the earth's gravitational force on you and points in the opposite direction.

Thus, through the gravitational interaction, you pull upward on the earth. I still marvel at this hidden force, revealed by applying the third law. Who would have thought that a mere human could pull the mighty earth?

To summarize this long answer to the triangle question: (1) Forces belong to interactions. (2) The third-law counterpart force of the gravitational force on you is $\mathbf{F}_{\text{you on earth}}$, which is the gravitational force on the earth *from* you. The counterpart is not $\mathbf{F}_{\text{ground on you}}$, which belongs to an entirely different interaction. Interaction was Newton's own view of the third law [2, pp. 568–569]:

> For all action is mutual... It is not one action by which the Sun attracts Jupiter, and another by which Jupiter attracts the Sun; but it is one action by which the Sun and Jupiter mutually endeavor to come nearer together (by the Third Law of Motion).

Rather than "action–reaction" with its easy causal-sequence misinterpretation, "interaction" is the heart of the third law. The distinction is illustrated in the following dialogue shared with me by Joshua Roth from his many years of teaching physics in Arlington High School in Massachusetts. A student, giving an example of a third-law pair, had suggested the causal sequence: "Action: I punch you in the face. Reaction: You punch me back." Roth: "No! That's the Mosaic law and maybe justice, but it's not Newton's third law. Action: You punch me in the face. Reaction according to Newton's law: You break your wrist." Or: "[W]e cannot touch without being touched" [8, p. 81].

1.2 Classifying Forces

Force is the star of the Newtonian drama. You can help yourself understand the play by enriching your force vocabulary. Therefore, we next discuss three ways to classify forces: as one of four fundamental interactions (Section 1.2.1), as active or passive (Section 1.2.2), and as short or long range (Section 1.2.3).

This seemingly tedious classification process may raise in you the following question marked with a leftward-pointing triangle. As I mention in the Preface (on p. ix), that triangle indicates a question that students ask or should ask me. Make a decent stab at an answer and compare your answer with the full explanation that follows.

◅ *Why go through all this effort to classify forces?*

I offer you an analogy from my learning piano. I just couldn't learn a difficult passage, the last line of Händel's Gavotte in G (Figure 1.2). My teacher showed me several ways to play and think about that line: by connecting the notes in the left hand into groups of notes (chords) while playing the right hand's melody, by connecting the right hand's notes while playing the left hand's melody, and by singing the right hand's melody while playing the left hand. After I tried these approaches, the line came to make sense, and I could play it as written. You are playing one of the hardest passages in physics, the concept of force, so take aid from and find comfort in all ways of reflecting on forces!

Figure 1.2 The last line of Händel's Gavotte in G (HWV 491). The right hand plays the notes on upper staff, and the left hand plays the notes on the lower staff.

1.2.1 The Four Interactions in Nature

Nature, as far as is known to science, uses four types of interactions and therefore four types of forces.

1. *Gravitational interaction.* This interaction acts between any two bodies, anywhere in the universe.

2. *Electromagnetic interaction.* This interaction acts only between charged bodies. It includes the electrostatic interaction between two charges and the magnetic interaction between moving charges (including magnets).

3. *Strong nuclear interaction.* This interaction acts between and holds together the quarks that make up protons and neutrons.

4. *Weak nuclear interaction.* This interaction acts between protons, neutrons, and electrons and can turn protons into neutrons and vice versa. It lies behind radioactive decay and nuclear fission. As its name suggests, it's much weaker than the strong nuclear interaction.

The last two interactions, the strong and weak nuclear interactions, have a minuscule range, about 1 femtometer (10^{-15} meters) or roughly the size of a nucleus. In the world around us, they have no direct effect. (Their indirect effect, however, is essential: Without them, protons and neutrons would not hold together, and there would be no atoms.) So, in learning and when using Newton's laws, we can ignore them.

With that simplification, classify every interaction – and the two forces that constitute it – as either gravitational or electromagnetic. The consequence:

If a force is neither gravitational nor electromagnetic, it doesn't exist.

Gravitational forces and interactions, because they join every pair of bodies in the universe, are surprisingly generic. In contrast, electromagnetic forces are diverse. They include contact forces between touching bodies, covalent bonds within molecules (for example, between hydrogen and oxygen atoms in water), hydrogen bonds between polar molecules (for example, between water molecules), ionic bonds within solids (for example, between sodium and chlorine ions in table salt), and Van der Waals bonds between nonpolar atoms or molecules (for example, between helium atoms or nitrogen molecules).

The classification into gravitational, electromagnetic, or nothing helps prevent a common and dangerous misconception.

➤ *Imagine a passenger sitting in a car or train wagon going around a turn. Is there an outward force on the passenger? If so, what's it called?*

You might suspect that there is an outward force and, if you had a proper education in Latin, that it's called the centrifugal force ("centrifugal" means away from the center). As a student, even lacking Latin, I would have agreed.

However, now you and I both know how to classify forces and interactions into one of four types. How then fares the alleged centrifugal force? Is it one of the two nuclear forces, either strong or weak? No. For as I mentioned on p. 5, no force in everyday life is a nuclear force. These forces act over too short a range (the size of a nucleus). Because "too short a range" is always the answer to a question about their relevance, I now really forget about the nuclear forces for the rest of the book.

Is the outward force a gravitational force? No. For no planet lies outside the vehicle's door and pulls the passenger toward it. Is it an electromagnetic force? If it is, what charges (or magnets) would be responsible for it? Perhaps they are in the door of the vehicle? But have you ever felt the door of a vehicle pulling you outward and toward it? I haven't. The door can only push the passenger inward and away from it.

Perhaps, instead, the outward force is due to the seat – that is, the seat acts on the passenger with an outward force. But how could the seat manage this feat? Its deformed part is its outward end, which gets compressed and thus pushes the passenger inward rather than outward. The mistake here, discussed further in Section 1.5.1, is confusing the force of the seat *on the passenger* (an inward force) with the force of the passenger *on the seat* (an outward force).

Indeed, and in answer to the triangle question, *no outward force acts on the passenger.* Thus, the alleged centrifugal force is none of the four forces in nature. It does not exist. (Why then do the passenger and vehicle move in a circle? This deep question, which involves all of Newton's laws, gets the longer answer that it deserves in Section 7.3.2.)

As a useful rule, do not mention the centrifugal force! Like most rules, it has an exception, discussed in Section 8.1. Until then, keep to the rule, avoid a widespread source of confusion, and greatly increase your chances of using Newton's laws correctly.

1.2.2 Active versus Passive Forces

The second force classification is into active versus passive forces. This classification, unlike the classification into four fundamental kinds of force (Section 1.2.1), isn't inherent in nature. Rather, it's a human choice.

1. *Active forces* are known gravitational and electromagnetic forces and pushes and pulls made intentionally by an animate being (be it a person, raccoon, or bear). Examples include the gravitational force on you or me and my push on a heavy box that remains sitting on the floor.

2. *Passive forces*, in contrast, arise and adjust themselves in response to active forces. One example is the force of the ground on you while you stand on the ground (Section 1.1). This force adjusts itself in response to the gravitational force on you as that active force tries to pull you into the ground. A second example is the force of friction preventing my moving that heavy box. This force adjusts itself in response to my active force on the box. As I push harder, the friction force grows in magnitude. If I am strong enough, the friction force can no longer adjust itself to match, and the box starts moving. This change illustrates a characteristic of passive forces, that they can adjust themselves in magnitude or direction only within limits.

This classification, like most human-created ones, isn't airtight. For example, when I hold a book over my head, the active force on the book is the downward gravitational force on it, but what kind of force is my upward force on it? Is it active because I'm animate? Or is it passive because I adjust how hard I push based on the gravitational force (pushing harder on a bigger book)? I would call it a passive force, but either choice fulfills the main purpose of this classification, which is to prevent us from overlooking a force from an inanimate object (like a door or a floor).

A puzzling feature of passive forces, especially given that they are produced (almost always) by inanimate object, is how they know their strength. How, for example, does the friction force on that heavy box (p. 7) adjust itself to prevent the box from moving even as I, frustrated by the box's obstinacy, increase how hard I push? This deep question implicates three subtle ideas: spring forces (Section 1.3.2), Newton's second law (Chapter 4), and acceleration (Section 6.2). Thus, its answer comes after their development (Section 7.1.6).

Until then, keep in mind the main reason for the idea of passive forces. It gives us a category, and therefore a name, for the forces exerted by inanimate objects. This category reminds us that inanimate objects can exert forces. These forces then become harder to overlook.

As a second benefit, the classification names an important connection between forces, one otherwise easily mislabeled as Newton's third law. When I stand on the ground and the gravitational force – here, the action and the active force – pulls me downward, the ground reacts by pushing me upward. This reaction, as you learned in Section 1.1, is *not* a third-law counterpart to the action (which is why I loathe the action–reaction name for the third law). However, this reaction *is* the passive force arising in response to the active force. When one force leads to another, the two forces cannot constitute a third-law pair; almost always, they comprise an active force and its corresponding passive force.

1.2.3 Long-Range versus Short-Range Forces

The third and final classification is into long-range forces, also known as body or volume forces, versus short-range forces, also known as contact or surface forces. For gravitational forces, this classification is easy. They are always long range: Gravitational forces are so weak that they require large sources, such as asteroids, moons, or planets, to have significant effects.

In contrast, electromagnetic forces, being relatively strong, don't require large sources of charge to have an appreciable effect. Thus, they can be either short or long range. One long-range example is the electromagnetic force on electrons in your retina due to jiggling electrons in the sun; that force is how you see the sun. Another example is the force on a compass needle due to electrons circulating in the earth's core. In contrast, the force between my finger and a pen, though also electromagnetic, is a short-range force (a contact force). As you learn when you meet spring forces (Section 1.3.2), this contact force is due to electrostatic repulsion between the outermost electrons in my skin cell's molecules and the outermost electrons in the pen's molecules.

In this book, with its focus on mechanics rather than electromagnetism, all electromagnetic forces will be short-range, contact forces. Thus, the classification into gravitational and electromagnetic forces will parallel the classification into long-range and short-range forces.

Either classification prevents you from overlooking forces. The short-range forces acting on a body are usually easy to spot: one due to each touching body. But the long-range forces are easier to overlook: out of sight, out of mind. By asking yourself, "On this body, are there also any long-range forces?" you are more likely to spot them too.

1.3 Important Forces

Understanding that forces belong to interactions (Section 1.1) is important, as is classifying forces (Section 1.2). However, you need forces to classify. Thus, you next meet the most important forces in the world around us. They, through Newton's laws, explain the motion of most everyday and heavenly bodies.

1.3.1 Gravitational Forces

The gravitational force comes in two seemingly different forms. One form is more famous: Newton's law of universal gravitation. It states that the gravitational force between mass m_1 and mass m_2 has magnitude

$$F = \frac{Gm_1 m_2}{r^2},\qquad(1.2)$$

where G is Newton's constant of gravitation, and r is the distance between the two masses (assumed to be points). As for the direction: The force is attractive, pointing from each mass to the other (Figure 1.3).

Specifying a vector, such as force, requires giving its magnitude and direction. Thus, never say that the gravitational force is $Gm_1 m_2/r^2$ – which provides only the magnitude. I recommend fanaticism about the distinction between a vector and its magnitude: Confusing these two quantities leads to many further difficulties with Newton's laws, an already subtle subject. (Even fanatics stray by mistake, so let me know if you find any such mistakes in this book.)

Figure 1.3 Gravitational interaction. Two particles, m_1 and m_2, separated by a distance r, participate in a gravitational interaction. The interaction's two forces have magnitude Gm_1m_2/r^2 and are opposite in direction.

If the bodies are not point particles but are spheres with a spherically symmetric density (which is roughly true for most planets), the force law (1.2) still works as long as r is measured between the bodies' centers (Newton invented calculus partly to prove this statement). In other words, a spherically symmetric body acts, for the purposes of the gravitational interaction, as if all its mass were concentrated at its center.

The most important gravitational force, at least for humanity, is the force of the sun on the earth. You could also argue for the gravitational force of the earth on each of us. However, without the sun holding the earth at just the right distance from the sun, giving the earth's surface just the right temperature to support life, none of us would be alive to argue for this alternative.

> *Roughly how large is the gravitational force of the sun on the earth?*

To find out, just put the appropriate values into the force law (1.2). Newton's constant (G) is roughly 7×10^{-11} crazy SI units, which are meters cubed per kilogram second squared ($\mathrm{kg^{-1}\,m^3\,s^{-2}}$). The earth's mass ($m_1$) is approximately 6×10^{24} kilograms. The sun's mass (m_2) is almost exactly 2×10^{30} kilograms. And the earth–sun distance (r) is roughly 1.5×10^{11} meters. Then

$$F \approx \frac{\overbrace{7 \times 10^{-11}\,\mathrm{kg^{-1}\,m^3\,s^{-2}}}^{G} \times \overbrace{6 \times 10^{24}\,\mathrm{kg}}^{m_1} \times \overbrace{2 \times 10^{30}\,\mathrm{kg}}^{m_2}}{\underbrace{\left(1.5 \times 10^{11}\,\mathrm{m}\right)^2}_{r^2}}. \qquad (1.3)$$

Rather than calculating F by breaking out the calculator, which would give us many spurious decimal places of precision and atrophy our intuitive sense for quantities in the world, let's estimate F by hand. Such calculations are best broken into three stages ordered from most to least important: the units, the powers of 10, and the mantissa (the remaining factor). The three stages are then reassembled to form the estimate:

$$F \approx \text{mantissa} \times 10^{\text{exponent}} \text{ units.} \qquad (1.4)$$

1. *Units.* This stage comes first because using the wrong units with even the correct number in front is dangerously wrong (as the sad crash of NASA's Mars Climate Orbiter shows [15]). Here, the units portion is

$$\frac{\overbrace{G}^{} \quad \overbrace{m_1}^{} \quad \overbrace{m_2}^{}}{\underbrace{kg^{-1}\, m^3\, s^{-2}}_{} \times \underbrace{kg}_{} \times \underbrace{kg}_{}}{\underbrace{m^2}_{r^2}}. \tag{1.5}$$

In the numerator, the kg^{-1} and one of the two kg factors cancel. Furthermore, the m^2 in the denominator cancels the m^2 within the numerator's m^3. So:

$$\frac{\cancel{kg^{-1}}\, m^{\cancel{3}}\, s^{-2} \times \cancel{kg} \times kg}{\cancel{m^2}} = \underbrace{kg\, m\, s^{-2}}_{N}. \tag{1.6}$$

The unit combination $kg\, m\, s^{-2}$ is the expanded form of the newton, abbreviated N, when written in terms of the fundamental SI units of kilograms, meters, and seconds: A newton *is* a kilogram meter per second squared. (Thus, force has dimensions of mass times length per time squared.) The force calculation so far, after this units stage, is then

$$F \approx \text{mantissa} \times 10^{\text{exponent}}\, N. \tag{1.7}$$

This units calculation has a useful side benefit. By starting with crazy units for G (namely $kg^{-1}\, m^3\, s^{-2}$) and ending with the correct units of force (namely newtons), it confirms the crazy units for G.

2. *Powers of 10.* Here, the powers of 10 are

$$\frac{\overbrace{10^{-11}}^{G} \times \overbrace{10^{24}}^{m_1} \times \overbrace{10^{30}}^{m_2}}{\underbrace{\left(10^{11}\right)^2}_{r^2}}. \tag{1.8}$$

The numerator contributes $-11 + 24 + 30$, or 43, powers of 10. The denominator contains 2×11, or 22, powers of 10. Dividing the numerator by the denominator gives $43 - 22$, or 21, powers of 10.

Thus, the force calculation so far, after this powers-of-10 stage, is

$$F \approx \text{mantissa} \times 10^{21}\, N. \tag{1.9}$$

3. *Mantissa (remaining factors).* After we set aside the units and the powers of 10, the following factors remain:

$$\frac{\overbrace{7}^{G} \times \overbrace{6}^{m_1} \times \overbrace{2}^{m_2}}{\underbrace{1.5^2}_{r^2}}. \tag{1.10}$$

The factor of 6 in the numerator combines with one factor of 1.5 in the denominator to give $6/1.5$ or 4. The factors of 7 and 2 in the numerator combine with the other factor of 1.5 in the denominator to give $14/1.5$ or almost exactly 10. Thus, the mantissa is approximately 4×10 or 40.

To summarize this long answer to the triangle question: The result of the three stages is that the magnitude of the gravitational force is

$$F \approx 40 \times 10^{21} \, \text{N} = 4 \times 10^{22} \, \text{N}. \tag{1.11}$$

The force points from the earth to the sun (Figure 1.4). The force of the earth on the sun has the same magnitude but points from the sun to the earth. (You can find further three-stage calculations in *The Art of Insight in Science and Engineering* [14, pp. 222, 307, and 336–337].)

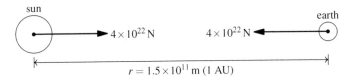

sun   earth

$4 \times 10^{22} \, \text{N}$ $4 \times 10^{22} \, \text{N}$

$r = 1.5 \times 10^{11} \, \text{m}$ (1 AU)

Figure 1.4 Earth–sun gravitational interaction. The two gravitational forces have the same magnitude, 4×10^{22} newtons, and are opposite in direction. (The separation is 1 astronomical unit (AU), and its value in meters is worth memorizing.)

Interestingly, although the earth orbits the sun because of the gravitational force of the sun, the earth never moves along the direction of this force. (It moves perpendicularly to the force.) This example illustrates the great subtlety of Newton's laws: that force does not produce motion itself. Motion can happen without any force. The great message of the second law, elaborated in Chapter 7, is that force only *changes* motion.

In contrast to (1.2), the second form of the gravitational force uses the constant g (lowercase!), known as the acceleration due to gravity or the gravitational acceleration. Its value is approximately 10 meters per second squared (m s^{-2}). In this form, the gravitational force on an object of mass m has magnitude

$$F = mg. \tag{1.12}$$

On me ($m \approx 60$ kilograms), and using the newton's definition (1.6),

$$F \approx 60 \, \text{kg} \times 10 \, \text{m s}^{-2} = 600 \, \text{kg m s}^{-2} = 600 \, \text{N}. \tag{1.13}$$

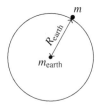

Figure 1.5 A body of mass m on the earth's surface. The earth acts like a point mass concentrated at the earth's center.

This second form only looks like a new form. As we'll soon see, it's just a useful approximation to the first form, Newton's law of universal gravitation (1.2), valid for bodies on or near the earth's surface (Figure 1.5). In that limit, the distance between the two bodies, r in (1.2), is approximately the earth's radius:

$$r \approx R_{earth} \approx 6.4 \times 10^6 \text{ m.} \tag{1.14}$$

The reason is that the earth has a spherically symmetric mass distribution (the case that I mentioned on p. 10). It therefore acts gravitationally like a point mass concentrated at the center of the earth.

Using the first form (1.2) with the separation r set equal to R_{earth}, the magnitude of the gravitational force on a body of mass m is

$$F = \left(\frac{Gm_{earth}}{R^2_{earth}} \right) m. \tag{1.15}$$

The parenthesized quantity, Gm_{earth}/R^2_{earth}, is the same for all bodies, so you can compute it just once. As a fairly accurate estimate,

$$\frac{Gm_{earth}}{R^2_{earth}} \approx \frac{7 \times 10^{-11} \text{ kg}^{-1} \text{ m}^3 \text{ s}^{-2} \times 6 \times 10^{24} \text{ kg}}{(6.4 \times 10^6 \text{ m})^2} \approx \underbrace{10 \text{ m s}^{-2}}_{g}. \tag{1.16}$$

The result is g! (Making a three-stage estimate for g is the subject of Problem 1.3.) With g replacing the parenthesized quantity in (1.15), the gravitational force F indeed has magnitude mg.

Fair warning: The $F = mg$ form with g as calculated in (1.16) is valid only for a body near (or on) the earth's surface, where the approximation $r \approx R_{earth}$ is reasonably accurate – as it is whenever the body's distance from the earth's surface is small compared to the size of the earth (thousands of kilometers). At the top of the atmosphere, about 10 kilometers up, g is only 0.3 percent smaller than it is at sea level. Even high above the atmosphere in low-earth orbit, g is only a few percent smaller than it is at sea level (as you show in Problem 1.5).

Even when the change in g is too large because the body is too high, you can often still use the $F = mg$ form, just with a modified g. If the body isn't *changing* its altitude h significantly, you just work out a modified g using (1.16) with R_{earth} replaced by $R_{earth} + h$. For example, for a satellite in a geosynchronous (a 24-hour) orbit, $h \approx 5.6R_{earth}$. At that height, g is a factor of 6.6^2 smaller than it is at sea level. However, as long as the satellite's altitude remains close to $5.6R_{earth}$, you can use $F = mg$ with the smaller g.

Back on earth, the $F = mg$ form (1.12) provides an enjoyable way to feel and remember the size of a newton. For it's said, perhaps apocryphally, that Newton, having fled Cambridge because of the plague, once sat under an apple tree in his garden in the English countryside, puzzling over the mysteries of gravity (many still unresolved). An apple fell from the tree onto his head and jolted him into the great insight that the motion of the apple and the motion of the moon, seemingly so different, can be explained by one, universal law of nature.

Figure 1.6 A small apple, perhaps the one that allegedly fell onto Newton. Its mass is approximately 0.1 kilograms, so mg is approximately 1 newton.

An apple in Newton's time, before Mendel, selective breeding, and genetic engineering, was small (Figure 1.6). The gravitational force on a small Newtonian apple, with a mass of roughly 100 grams or 0.1 kilograms, has a magnitude of about 1 newton:

$$F \approx \underbrace{0.1 \, \text{kg}}_{m} \times \underbrace{10 \, \text{m} \, \text{s}^{-2}}_{g} = 1 \, \text{N}. \tag{1.17}$$

1.3.2 Spring Forces

Once upon a time, the word "spring" reminded me only of the large metal coils under a bicycle seat or car frame that make the ride smoother. The forces exerted by such macroscopic springs, however, originate in microscopic spring interactions – in the short-range electromagnetic interactions between atoms or molecules. These interactions, like macroscopic spring interactions, can be attractive (when a spring is stretched) or repulsive (when a spring is compressed). To see how atoms (or molecules) can generate attraction and compression, consider as an example a crude model of sodium chloride, table salt. A sodium atom easily gives up its lone and weakly bound electron in its outermost shell. A neighboring chlorine atom easily accepts this gift and thereby completes its outermost electron shell (technically, its $3p$ shell). The gift and its acceptance create an interaction between the positively charged sodium ion Na^+ and the negatively charged chlorine ion Cl^- (also called chloride).

$$Na^+ \quad \underset{\text{interaction}}{\overset{\text{attractive electromagnetic}}{- - - - - - - - - - - - - - - - -}} \quad Cl^-$$

Figure 1.7 Electromagnetic attraction between distant sodium and chloride ions.

When these ions are far apart, they see each other as shown in Figure 1.7, as a single positive facing a single negative charge. The interaction between these opposite charges is attractive – as if a spring between them were stretched.

When the ions are close to each other, they see each other's internal structure: positively charged protons in the nucleus and negatively charged electrons surrounding the nucleus. Now the strongest interaction is between the closest charges, which are in the electron clouds. This electromagnetic interaction between like charges is repulsive (Figure 1.8). Thus, at close range, the ions repel each other – as if a spring between them were compressed.

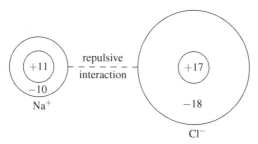

Figure 1.8 Electromagnetic repulsion between close sodium and chloride ions.

Somewhere in between, the bond between the sodium ion and the chloride ion has its natural, or relaxed, length where the interaction is neither repulsive ("neither too close" in Goldilocks's terms) nor attractive ("nor too far"). This property isn't specific to sodium chloride but applies to any interatomic interaction.

Thus, substances have a natural size and shape, when all their interatomic bonds have their natural lengths. Deviations from the natural state bespeak an outside interaction. Imagine, for example, a water glass standing on a table. The glass participates in two interactions: a gravitational interaction with the earth and a spring interaction with the table. The spring interaction means that the chemical bonds near the surface of contact, in the glass and the table, change their lengths: The table and glass deform slightly. To make such deformations visible, I sit on a soft sofa and observe how the cushion and my rear end deform!

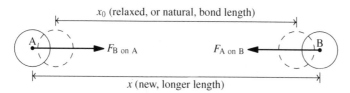

Figure 1.9 A stretched atomic-bond spring. Atoms A and B have a natural separation of x_0. When their separation x is greater than x_0, their spring interaction consists of two attractive forces.

When the change in length of a spring is small, whether a physical spring or an interatomic or intermolecular bond spring (Figure 1.9), the two forces in the spring interaction have a magnitude given by Hooke's law:

$$F = k|x - x_0|, \qquad (1.18)$$

where k is called the spring constant and measures the spring's stiffness, x_0 is the spring's natural or relaxed length, and $|x - x_0|$ is then the change in the spring's length (compared to its natural length). Thus, when the spring is compressed, $x - x_0$ is negative, but $|x - x_0|$ is still positive – keeping F (a magnitude) positive.

You may have seen Hooke's law as $F = -kx$. However, the more complex form (1.18) is better. It makes the natural length explicit, rather than hiding it in the meaning of x (as the spring extension). It also has no minus sign. Thus, it doesn't invite us to confuse the magnitude of a spring force, which is never negative, with the value of the force's component (which could be negative).

The two forces in the interaction oppose the change from the spring's natural length. In a coordinate system with the positive x axis pointing to the right (from atom A to atom B), the force *vectors* have the components

$$
\begin{aligned}
F_x^{A\ \text{on}\ B} &= -k(x - x_0); \\
F_x^{B\ \text{on}\ A} &= +k(x - x_0).
\end{aligned}
\qquad (1.19)
$$

The minus sign in the spring force's component $F_x^{A\ \text{on}\ B}$ means that, when the spring is stretched ($x > x_0$), the force on atom B fights the stretch and pulls atom B back toward atom A. The positive sign in $F_x^{B\ \text{on}\ A}$ has the same function, making the force on atom A also fight the stretch.

▶ *Do these formulas for the force components change when the spring is compressed (rather than stretched)?*

No! Figure 1.10 shows the compressed spring with the resulting forces and their directions. Let's check that the components are consistent with (1.19).

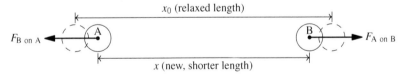

Figure 1.10 A compressed atomic-bond spring. Atoms A and B have a natural separation of x_0. When their separation x is less than x_0, their spring interaction consists of two repulsive forces.

In compression, $x - x_0$ is negative, so the force component from (1.19),

$$F_x^{A\ \text{on}\ B} = -k(x - x_0), \qquad (1.20)$$

is positive – the minus sign in front cancels the negative value of $x - x_0$ – as it should be. The compressed spring tries to push atom B away from atom A, in the positive x direction and back to the equilibrium position where the spring has its relaxed length.

Now imagine again the water glass standing on the table. The contact forces in the table–glass interaction are each the sum of zillions of tiny interatomic (or intermolecular) electromagnetic spring forces. In other words, contact forces are spring forces, even when the deformation of the springs is too tiny to see; and spring forces are short-range electromagnetic forces.

1.3.3 Drag

Perhaps a sign that we live in a fallen world, among the most prevalent forces in everyday life are friction forces. Thus, we now meet the most important friction forces, in order of increasing conceptual complexity: drag (this section), dynamic friction (Section 1.3.4), and static friction (Section 1.3.5).

Drag, also known as air or fluid resistance, opposes the motion of any object moving through a fluid. It results from contact between the object and fluid, so it's another short-range electromagnetic contact force. For most everyday objects in most fluids, its magnitude is given roughly by

$$F_{\text{drag}} \sim \rho v^2 A_{\text{cs}}, \tag{1.21}$$

where ρ is the fluid's density, v is the object's speed, and A_{cs} is the object's cross-sectional area (its area perpendicular to the flow). Figure 1.11, for example, shows a solid cone moving with speed v to the right and the same cone seen head on as it approaches you. The cross-sectional area, A_{cs}, is the area of the cone's back surface (the shaded area), not the area of the cone's whole surface. (Here and in subsequent figures, velocity is indicated with an arrow having only a single harpoon – a reminder that the arrow is not a force arrow.)

With that understanding, you can see that the F_{drag} formula (1.21) makes physical sense. Running in a swimming pool (high ρ) is much harder than running in air (low ρ), so F_{drag} should, and does, increase as ρ increases. Similarly, running rapidly in water (high v) is much harder than running slowly in water (low v), so F_{drag} should, and does, increase as v increases. Finally, bicycling upright (large A_{cs}) is harder than bicycling in a crouch (low A_{cs}), so F_{drag} should, and does, increase as A_{cs} increases.

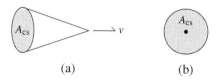

(a)   (b)

Figure 1.11 Cross-sectional area A_{cs}. (a) A cone moving in a fluid. (b) The same cone heading toward you. The shaded area that you see is its cross-sectional area.

The ~ ("twiddle" or "tilde") in (1.21) indicates that the relation omits a dimensionless constant, meaning a pure number such as 0.7, π, or 2.3. This missing number can be calculated in special circumstances and can be measured in a wind tunnel in many circumstances. As a useful rule of thumb, the missing number isn't too different from 1.

As an example of drag's importance, let's estimate the magnitudes of the two forces on a ping-pong ball (a hollow plastic ball an inch or a few centimeters in diameter): the gravitational force and the drag force. (The ball experiences a third force, buoyancy, but this force is tiny and ignored throughout this book.)

1. *Gravitational force.* A ping-pong ball has a mass m of approximately 10 grams (about one-third of an ounce) or 10^{-2} kilograms. The gravitational force acting on it, from (1.12), has magnitude mg. Thus,

$$F_{\text{g}} \approx \underbrace{10^{-2}\,\text{kg}}_{m} \times \underbrace{10\,\text{m s}^{-2}}_{g} = 0.1\,\text{N}. \qquad (1.22)$$

2. *Drag force.* The drag force depends on the ball's speed. Let's imagine that one player served the ball and that the second player has hit it back with decent speed (in my misspent youth, I played much ping-pong and could have been that second player). On the return flight, the ball's speed v is, say, 15 meters per second (33 miles per hour). Its cross-sectional area is about 5 square centimeters or 5×10^{-4} square meters. And the density of air is roughly 1 kilogram per cubic meter. With these values,

$$F_{\text{drag}} \sim \underbrace{1\,\text{kg m}^{-3}}_{\rho} \times \underbrace{\left(15\,\text{m s}^{-1}\right)^{2}}_{v^2} \times \underbrace{5 \times 10^{-4}\,\text{m}^2}_{A_{\text{cs}}} \approx 0.1\,\text{N}. \qquad (1.23)$$

The drag force has approximately the same magnitude as the gravitational force (Figure 1.12)! Thus, drag strongly affects how the ping-pong ball moves.

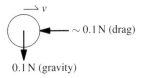

0.1 N (gravity)

Figure 1.12 Forces on the ping-pong ball. The ball, moving to the right at speed v, experiences a drag force to the left. It also experiences a gravitational force downward. For typical ping-pong speeds, the two forces are comparable in magnitude.

Drag is a cousin to two other forces due to motion in a fluid: lift and thrust. To describe these forces, and many others later in this book, I first need to distinguish a component (a standard term) from a portion – a term that I invented for this book in order to seal off conceptual traps.

A component of a vector is the vector's projection in a given direction: "The z component of the gravitational force is 20 newtons" (if the z axis points down). Symbolically, the z component F_z is defined as a dot product:

$$\underbrace{z \text{ component of a vector } \mathbf{F} \equiv \mathbf{F} \cdot \hat{\mathbf{z}},}_{F_z} \tag{1.24}$$

where the triple-equals sign ($\equiv$) means "is defined to be," and $\hat{\mathbf{z}}$ ("z hat") is the unit vector in the $+z$ direction. A component's value can depend on the direction chosen, but a component has no direction itself. Thus, it's not a vector.

In contrast, a portion is a piece of a vector: "The z portion of the contact force is 20 newtons downward; the x portion is 10 newtons to the left." A portion, being itself a vector, has magnitude and direction. The z portion $\mathbf{F}_z$ is defined as the z component times the $\hat{\mathbf{z}}$ unit vector (multiplication by the unit vector turns the component into a vector):

$$\underbrace{z \text{ portion of a vector } \mathbf{F} \equiv}_{\mathbf{F}_z} \underbrace{(\mathbf{F} \cdot \hat{\mathbf{z}}) \, \hat{\mathbf{z}}.}_{F_z} \tag{1.25}$$

Of particular relevance to drag and thrust, we can also speak of *an x* (or a *y* or a *vertical*, etc.) portion of an overall force, meaning a conceptually separate force in the x direction that contributes or belongs to the overall force.

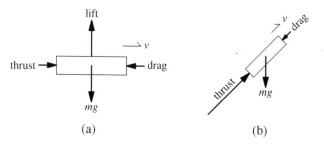

(a)   (b)

Figure 1.13 (a) The three portions of the contact force of the air on a plane: (1) Lift is the portion perpendicular to the plane's motion; (2) thrust is a portion along the plane's motion; (3) drag is a portion opposite to the plane's motion. (b) Thrust and drag on a rocket. Thrust is the contact force of the hot exploding fuel on the rocket. (The gravitational forces are included for completeness.)

A plane experiences lift, drag, and thrust – all portions of one overall force, the electromagnetic contact force of the air on the plane (Figure 1.13a). Lift is the portion perpendicular to the plane's motion. Thrust is a portion along the plane's motion. And drag is a portion opposite to the plane's motion. (Thus, *the portion parallel to the plane's motion is the sum of thrust and drag.*) Drag is the force attributable purely to the plane's motion, without the engines running. Thrust is the force attributable to the running engines. However, beware of a common trap: The engines don't supply the thrust, even though we speak that way informally. Rather, they move the air in such a way that the air supplies a forward force on the plane. Thrust is a portion of the contact force of the air.

A rocket (Figure 1.13b) experiences thrust and drag (but usually not lift). Drag is the contact force of the air or, when the rocket is in space, of the interplanetary gas. Thrust, in contrast to its interpretation for a plane, is the contact force of the hot exploding fuel. The repulsive fuel–rocket contact interaction has two equal-and-opposite sides: (1) the rocket exerting a backward force on the fuel, and (2) the fuel exerting a forward force – the thrust – on the rocket.

Failure to understand this application of Newton's third law led even knowledgeable contemporaries of Robert Goddard in the 1920s to ridicule his ideas for rocket flight [6]. They insisted that rocket flight was impossible without air "to push against." I write these words on the 50th anniversary of Neil Armstrong's walk on the moon; Goddard and the third law get the last laugh.

1.3.4 Dynamic Friction

When one body touches and slides past another, each experiences a contact force – a short-range electromagnetic force. This force can be divided into two portions. The portion parallel to the surface of contact is called dynamic or sliding friction – our second important friction force. The portion perpendicular to the surface of the contact is called the normal force. Here, "normal" is used in its mostly archaic meaning of "perpendicular."

Dynamic friction's magnitude F_μ is given by the empirical relation

$$F_\mu = \mu N, \tag{1.26}$$

where μ is the sliding-friction coefficient (it's a dimensionless constant like 0.83 or 0.02), and N is the magnitude of the normal force. (A coefficient means "a dimensionless quantity that we cannot calculate from first principles, so we measure it or look it up in a table.")

A slippery surface means a low μ. For example, for ice skates sliding on recently groomed and smooth ice, μ is about 0.001. When I skate, a rare event because I easily fall over, the normal force has roughly the same magnitude as the gravitational force on me, whose magnitude is 600 newtons. (The two magnitudes are equal based on Newton's second law, as you learn in Section 5.1.) So the dynamic or sliding friction on me is (in magnitude) about 0.6 newtons:

$$F_\mu \approx \underbrace{0.001}_{\mu} \times \underbrace{600\,\text{N}}_{N} = 0.6\,\text{N}. \tag{1.27}$$

Dynamic friction – like any interaction or part of an interaction – results in two forces: (1) a 0.6-newton force on me that points backward (opposite to my direction of motion), and (2) a 0.6-newton force on the ice that points forward (in my direction of motion). By skating, I try to pull the ice forward – though with little success as the ice is connected to the massive earth.

1.3.5 Static Friction

The third important "friction" is static friction. Like dynamic friction, it occurs when two bodies are in contact and is the parallel portion of the contact force. Its magnitude F_{static} also looks similar to dynamic friction's magnitude (1.26):

$$F_{\text{static}} \leq \mu_s N, \tag{1.28}$$

where μ_s is the static-friction coefficient, and N is again the magnitude of the normal force (the perpendicular portion of the contact force). Because of this structural similarity, static friction is misleadingly called a friction force. I will discuss why the name "static friction" is misleading after I unpack the magnitude equation (1.28).

The interesting features of (1.28) are the $\leq$ sign and the absence of information on the force's direction. The reason for both features is that static friction is a passive force (Section 1.2.2): It adjusts itself to *prevent* two bodies from moving relative to each other. (The seemingly magical adjustment process is demystified in Section 7.1.6.) It's therefore the enemy of dynamic friction, trying to prevent dynamic friction from even occurring.

Imagine a box sitting on a hill (Figure 1.14). The box is held in place by static friction, which points uphill. This force adjusts its strength and direction in order to hold the box in place. When you push the box uphill without moving it, static friction reduces its magnitude. If you push hard enough, static friction changes direction and points downhill, in order to hold the box in place.

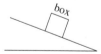

Figure 1.14 A box sitting on a hill and held in place by static friction.

Now place the box on level ground and try gently but unsuccessfully to push it forward. As long as the box doesn't move, your push is opposed by static friction, which points backward. If you instead push the box backward (and the box still doesn't move), static friction again opposes your push – by pointing forward. Not only the magnitude but also the direction of static friction adjusts to keep two bodies from moving relative to each other.

However, this adjustment happens only within limits, as indicated by the $\leq$ sign in the magnitude (1.28). Beyond a certain limit, static friction gives up. That limit depends on the normal force and the static-friction coefficient μ_s. If you oil the hill – thereby reducing μ_s and the static-friction limit – the box will probably slide downhill (and become subject to the enemy, dynamic friction).

As I mentioned at the start of this section, the name "friction" is badly chosen for static friction. For friction contains the ideas of energy loss, of opposing a body's motion, and of turning the energy of this motion into heat energy. There-fore, air drag, often called air friction, is a legitimate friction. However, static friction, unlike dynamic (sliding) friction, involves no conversion of motion energy into heat. Nor can it, as the two touching bodies have no relative mo-tion: the whole point of the adjective "static." In short, "static friction" is an oxymoron (but the name is too entrenched to change).

1.4 Force Magnitudes

You have now met the forces essential for applying Newton's laws to the world around you. To give you a feel for these forces' strengths and for the size of the newton, we'll estimate the magnitudes of diverse forces and place these magnitudes on a logarithmic scale. In contrast to a linear scale, on which a given distance corresponds to a particular difference, on a logarithmic scale a given distance corresponds to a particular ratio. A single logarithmic scale can thereby show forces or interactions of vastly different strengths (Figure 1.15).

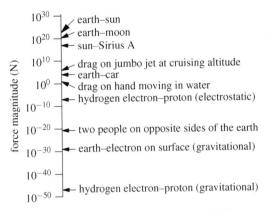

Figure 1.15 Force magnitudes placed on a logarithmic scale. Many are explained in the text, and a few are left for you (Problem 1.8).

The weakest interaction that we'll place is the gravitational interaction between the proton and electron in a hydrogen atom. From (1.2),

$$F_g \approx \frac{\overbrace{7 \times 10^{-11}\ \mathrm{kg}^{-1}\ \mathrm{m}^3\ \mathrm{s}^{-2}}^{G} \times \overbrace{1.7 \times 10^{-27}\ \mathrm{kg}}^{m_1} \times \overbrace{10^{-30}\ \mathrm{kg}}^{m_2}}{\underbrace{\left(0.5 \times 10^{-10}\ \mathrm{m}\right)^2}_{\text{hydrogen's radius}}} \tag{1.29}$$

$$\approx 5 \times 10^{-47}\ \mathrm{N}.$$

From Newton's third law, this value is the strength of either side of the interaction: of the gravitational force on the electron due to the proton or of the gravitational force of the proton due to the electron.

Much stronger is the electrostatic interaction between the same electron and proton (electrostatics is the particular case of an electromagnetic interaction with no magnetic field and fixed electric fields). Its strength is given by Coulomb's law of electrostatics:

$$F_e = \frac{1}{4\pi\epsilon_0} \frac{|q_1 q_2|}{r^2}, \tag{1.30}$$

where ϵ_0 is the permittivity of free space, q_1 and q_2 are the two charges, and r is the distance between the charges. Like Newton's law of universal gravitation (1.2), Coulomb's law is also an inverse-square law, meaning that the interaction's strength is proportional to $1/(\text{separation distance})^2$. Unlike gravitation, which is always attractive, the electrostatic interaction can be repulsive or attractive, depending on whether the charges are, respectively, like or unlike.

Putting in the electron and proton charges and hydrogen's radius a_0,

$$F_e \approx \frac{1}{4\pi\epsilon_0} \frac{\overbrace{\left(1.6 \times 10^{-19}\ \mathrm{C}\right)^2}^{|q_{\text{electron}} q_{\text{proton}}|}}{\underbrace{\left(0.5 \times 10^{-10}\ \mathrm{m}\right)^2}_{a_0^2}} \approx 10^{-7}\ \mathrm{N}, \tag{1.31}$$

where C is the abbreviation for the charge unit of the coulomb.

This interaction is more than a factor of 10^{39} stronger than the gravitational interaction (1.29) between the same electron and proton! (Because Coulomb's law and Newton's law of gravitation are both inverse-square force laws, this gigantic ratio applies to *any* electron–proton separation.)

At the strongest end of our scale sits the gravitational interaction between the earth and sun. Its strength was calculated in three stages to reach (1.11).

In between sit forces of everyday experience. One example is the gravitational force on a car. A medium-sized car has a mass of roughly 1000 kilograms. Based on the approximation (1.12), that $F = mg$ near the earth's surface, the gravitational force on the car is roughly 10^4 newtons.

Another everyday force is the drag force on a hand moving through water. To feel it yourself, fill a kitchen sink with water. Then drag (sorry!) your open, slightly cupped hand through the water at a moderate speed, roughly 1 meter per second. If your cupped hand's cross-sectional area is roughly 10^{-2} square meters (10 centimeters by 10 centimeters), then the drag force's magnitude (1.21) is roughly 10 newtons:

$$F_d \sim \underbrace{10^3 \,\text{kg}\,\text{m}^{-3}}_{\rho_{\text{water}}} \times \underbrace{\left(1\,\text{m}\,\text{s}^{-1}\right)^2}_{v^2} \times \underbrace{10^{-2}\,\text{m}^2}_{A_{cs}} = 10\,\text{N}. \qquad (1.32)$$

I just tried the experiment. Moving my hand through the water felt as effortful as holding up a medium-sized book against gravity. And the gravitational force on a medium-sized book – say, a 1-kilogram mass – is also roughly 10 newtons, which supports the estimate (1.32) of the drag-force magnitude.

1.5 Forces to Avoid

Contrasting the essential forces of Section 1.3 are two terrible forces: the centripetal force and the centrifugal force. When using Newton's laws, these forces usually lead to disaster. Thus, never use them. If this bare assertion is sufficient for your purposes, you could skip the rest of this section. If you would like to know the reason for my dogmatism and what problems it prevents, read on.

1.5.1 Centrifugal Force

"Centrifugal" is a Latin compound meaning "away from the center." Thus, a centrifugal force is an outward force. It's typically invoked for a body moving in a circle – for example, on the passenger of Section 1.2.1 sitting in a car or train going around a turn. The alleged centrifugal force is directed outward, away from the center of the circle containing the turn. As I hopefully convinced you in Section 1.2.1 by considering the four kinds of interaction in nature, the centrifugal force isn't an actual force and should not be invoked.

Yet it's invoked often anyway, for two reasons. First, an outward force often exists. This force is applied *by* the passenger, rather than to the passenger. When the vehicle rounds that turn, the passenger pushes outward against the vehicle's door (or the seat belt). Now two related but distinct meanings of "push" cause a problem. In physics, "to push [on an object]" means simply "to exert a force that points into an object." In everyday life, however, "to push" implies an agent acting with intent. For example, a book on a table, although it pushes on the table in the physics sense, is rarely in everyday life said to push on the table.

Thus, in the everyday sense, the passenger, who pushes on the door only involuntarily, is *not* pushing on the door. The passenger is a passive body who, in a passive construction, "gets pushed." And a force conveniently, though incorrectly, makes itself available to do the pushing: the centrifugal force!

The medicine that treats this subtle malady is Newton's third law. The door and the passenger share a contact interaction. Because of the nature of their contact, which is mere touch without glue or other attractive bond, the interaction must be repulsive. Because of the relative positions (the door on the outside and the passenger on the inside), the interaction's forces are outward on the door and inward on the passenger. The centrifugal force is neither needed nor valid.

The second, valid reason for invoking the centrifugal force is that it's needed in a rotating reference frame. Such reference frames, along with their extra forces (of which the centrifugal force is one), are a powerful tool (and are touched upon in Section 8.1). But using them is like swinging a chainsaw: You can easily make terrible mistakes and cut off your leg. Until you can make beautiful furniture with hand tools, set down the chainsaw: Until you are skilled at using Newton's laws without rotating reference frames and their extra forces, don't mention the centrifugal force. It does not exist!

1.5.2 Centripetal Force

The mirror image of the centrifugal force is the centripetal force. "Centripetal" is a Latin compound meaning "toward the center." Thus, a centripetal force is a force toward the center, or inward. In comparison to the centrifugal force, the centripetal force can exist, making its overuse more tempting.

Consider two versions of the passenger in a train car rounding a turn.

1. *The seat is a perfectly smooth and frictionless bench, and the passenger also presses against a frictionless outer wall (for example, a door).* The passenger, who touches two bodies, experiences two short-range contact forces. The contact force of the bench is upward (without friction, the contact force has no horizontal portion). The contact force of the frictionless outer wall, similarly, has no vertical portion and points inward: It's a centripetal force. As the only such force, it's also *the* centripetal force. In this version, a centripetal force does exist, and its use is correct.

2. *The seat is rough enough that the passenger doesn't slide even without touching the wall.* Now the passenger experiences only one contact force, from the seat. Being the sum of the two contact forces in the first version, this contact force points upward and inward. The only other force acting on the passenger is the gravitational force, which points downward. Thus, no force on the passenger points inward, so no force is a centripetal force.

These versions illustrate two tricky aspects of the centripetal force. First, the centripetal force, even on a body moving in a circle, might not exist (version 2). Second, when it does exist (version 1), it's not a new kind of force. Rather, "centripetal" merely redescribes an actual, physical force (one of nature's four fundamental kinds in Section 1.2.1) that happens to point directly inward.

Thus, you have a choice. You can keep these two points in mind, invoking a centripetal force only when one exists and remembering that it's not a new force. Or you can simplify your life by abstaining from the centripetal force completely. I choose and recommend abstinence because I find Newton's laws and force subtle enough without adding avoidable complications.

In summary, forget the centripetal and the centrifugal forces.

1.6 Problems

1.1 For each of the following given forces, (i) determine the two interacting bodies A and B, (ii) write the force in the form $F_{A \text{ on } B}$ (replacing A and B by their short names in that situation), (iii) express the force's third-law counterpart force in symbols and in words, and (iv) give the counterpart force's direction. Here are the forces:
   a. the force of the ground on you when you stand on the ground,
   b. the gravitational force on a freely falling stone,
   c. the force of a tree branch on a cherry (and stem) attached to and hanging from the tree, and
   d. the lift force on a hummingbird as it hovers above a flower.

1.2 For each force given in Problem 1.1, classify the force as active or passive, electromagnetic or gravitational, and short range or long range. (Each force's third-law counterpart force will have identical classifications.)

1.3 Make the three-stage estimate (without a calculator!) – units, exponent, and then mantissa – for the g calculation (1.16).

1.4 Estimate the gravitational force of the moon on the earth (in magnitude). Feel free to look up the moon's mass and its distance from the earth (center to center), also known as its orbital radius. How does the force compare to the gravitational force of the sun on the earth (1.11)?

1.5 Calculate, to two or three decimal places, the dimensionless ratio

$$\frac{g \text{ at an altitude of } 200 \text{ kilometers (a low-earth orbit)}}{g \text{ at sea level}}. \quad (1.33)$$

Is the ratio close to what you expected? (I found it surprising.)

1.6 Estimate and compare the magnitudes of the drag force and the dynamic-friction force on an Olympic speed skater.

1.7 Return to the box sitting peacefully on the hill in Section 1.3.5 and to the static-friction force on the box. What's the Newton's-third-law counterpart to this force? In what direction does this counterpart force point?

1.8 Looking up needed masses and distances, confirm the approximate placement on the logarithmic scale (Figure 1.15) of the following forces and interactions:

   a. the gravitational interaction between the earth and an electron on the surface of the earth,

   b. the (gravitational) interaction between two people on opposite sides of the earth,

   c. the drag force on a jumbo jet at cruising speed and altitude, and

   d. the (gravitational) interaction between the sun and Sirius A (the brightest star in the night sky, also known as the Dog Star).

2

Freebody Diagrams: Representing Forces

Forces easily get overlooked, especially among several bodies sharing many interactions. The tool for tracking and representing forces is the freebody diagram: the analog of a circuit diagram. A circuit diagram makes a circuit comprehensible; a freebody diagram makes a mechanical system comprehensible.

2.1 Making Freebody Diagrams: A Foolproof Recipe

The name *freebody diagram* describes, in abbreviated form, how to make one. Each of the name's three parts – *free*, *body*, and *diagram* – carries an important meaning and commandment.

1. *Diagram.* The forces shall be represented pictorially, with arrows indicating their magnitudes and directions.

2. *Body.* The diagram shall show the forces acting on exactly one body, called the primary body. Thus, the diagram of you standing on the ground (Figure 1.1) shows the forces acting only on you, not on the ground. The body can be several bodies lumped mentally into one composite body: for example, you and your socks and shoes. We then care about only the forces from outside the composite body and ignore the forces within it (you learn why in Section 7.2).

3. *Free.* The body shall be drawn free of contact with other bodies. The example diagram represents only you, not the earth – even though you touch the earth.

These commandments are built into the following recipe for a correct freebody diagram. The recipe uses two principles: Interactions are short or long range (Section 1.2.3), and a force is one side of an interaction (Newton's third law).

> ***How to Make a Correct Freebody Diagram Every Time***
> 1. Handle the short-range, contact interactions.
>   1a. Free the primary body.
>   1b. Replace broken contacts with contact forces.
> 2. Handle the long-range interactions.
>   2a. Find the long-range interactions.
>   2b. Represent these interactions by drawing long-range forces.
> 3. Optional: Include a ghostly reminder of the forces' origins.

To illustrate the recipe, look again at the familiar situation of a person standing on the ground. Our goal is a freebody diagram of the person, the primary body.

1. *Handle the short-range, contact interactions.* Interactions can be short or long range. But handle the short-range interactions first because they are the easier ones to find and because doing so frees the primary body right away.

   1a. *Free the primary body.* Tear the primary body from the secondary bodies that touch it (Figure 2.1). The primary body thereby gets its own diagram free of other bodies – thereby satisfying commandment 3.

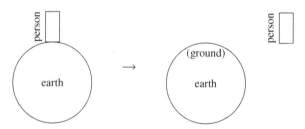

Figure 2.1 Step 1a: Freeing the primary body (the person). The "free" in freebody diagram means that the primary body is drawn separated from all other bodies.

   Here, the primary body is the person, and the torn-off (and only) secondary body is the earth. Making a secondary diagram, an optional step, gives us practice with Newton's third law, so let's make it for this example. (This secondary diagram, thanks to the rest of this recipe, might also become a freebody diagram, as it will here.)

   1b. *Replace broken contacts with contact forces.* Each tear breaks one contact interaction. Here, the tear breaks the interaction between the person and the ground (as a part of the earth). According to Newton's third law, an interaction has two equal and opposite sides. Here, these sides are $\mathbf{F}_{\text{person on ground}}$ and $\mathbf{F}_{\text{ground on person}}$.

   Place one side of each contact interaction – the force *on* the primary body – on the primary diagram (Figure 2.2). Place the other side – the force *due to* the primary body – on the appropriate secondary diagram (if it's being made). Here, you place and label $\mathbf{F}_{\text{ground on person}}$ on the person's diagram; and you place and label $\mathbf{F}_{\text{person on ground}}$ on the earth's diagram (because the ground belongs to the earth).

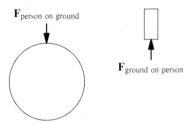

Figure 2.2 Step 1b: Replacing broken contact with contact forces.

As a notational simplification, you can omit the "on person" or "on ground" qualifiers in the force labels. Thus, on the person's diagram, you can label the force as simply $\mathbf{F}_{\text{ground}}$ – with the "on person" qualifier understood because it's the person's diagram. The shorter label is unambiguous thanks to step 1a when you freed the primary body.

The secondary diagrams give us a place to draw the opposite side of each contact force on the primary body. They remind us of the reciprocal nature of Newton's third law and prevent us from placing the opposite side on the primary diagram – a common fatal mistake.

As a reminder that contact forces act on a body's surface rather than throughout the body, draw these arrows outside but touching the body. When the contact interaction is repulsive, the more common case, *outside but touching* means that the arrow's tip touches the body (leaving the tail outside the body). When the contact interaction is attractive – say, between a schoolchild's desk and the piece of chewing gum stuck underneath the desk – then *outside but touching* means that the tail touches the body (leaving the tip outside the body).

Here, the person–ground interaction is repulsive – each body tries to repel the other – so $\mathbf{F}_{\text{ground on person}}$ is drawn with its tip along the person's bottom surface (and its tail outside the person).

2. *Handle the long-range interactions.* The other interactions are long range.

2a. *Find the long-range interactions.* Now identify all of the primary body's long-range interactions. These interactions are usually gravitational – but not always. Two bar magnets, for example, have a long-range electromagnetic interaction. In the example here, the sole long-range interaction is the gravitational interaction between the person and the earth.

2b. *Represent these interactions by drawing long-range forces.* As with the short-range interactions (step 1b), each long-range interaction has two sides. Place one side of each long-range interaction – the (labeled) force *on* the primary body – on the primary diagram (Figure 2.3). Place the other side – the (labeled) force *due to* the primary body – on the appropriate secondary diagram (if it's being made). Thus, place the gravitational force $\mathbf{F}_{\text{earth on person}}$ on the person's diagram; and place the gravitational force $\mathbf{F}_{\text{person on earth}}$ on the earth's diagram.

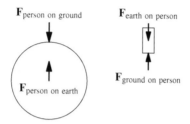

Figure 2.3 Step 2b: Drawing the long-range forces. After identifying the long-range interactions, draw the pair of forces from each interaction.

As a reminder that long-range or body forces act throughout the body rather than only on its surface, draw the long-range forces at least partly inside the body. But don't place either end (tip or tail) at the surface – to avoid hinting that the force is short range. Furthermore, for gravitational long-range forces – and, in mechanics, all long-range forces are gravitational – try to place the tail or tip at the body's center of mass. But this advice is only a rule of thumb. First, typesetting considerations sometimes override it. Second, "the" gravitational force is really the sum of many tiny gravitational forces acting on a body's many particles. Fortunately, in most analyses, they act as if they were combined into a single force acting at the body's center of mass.

What is a body's center of mass?

A body's center of mass is located at the weighted average of the locations of its individual particles, where each particle is weighted by its mass. Think of each particle voting for itself as the center of mass: "Me, me, me!" Each particle gets a number of votes proportional to its mass. The inevitable discord is resolved by putting the center of mass at the (vector) average. For example, when a body consists of two identical particles, its center of mass is midway between the particles. If one particle is twice as massive as the other, the center of mass is one-third of the way from the more to the less massive particle.

Now the primary body has a freebody diagram! When a secondary body interacts only with the primary body, then its diagram is also a freebody diagram. Here, the secondary body (the earth) interacts only with the person (through gravitational and contact interactions), so the earth's diagram is also a freebody diagram. (When the secondary body has other interactions and you need its freebody diagram, follow this recipe with the secondary body as the new primary body.)

3. *Optional: Include a ghostly reminder of the forces' origins.* As an improvement helpful in complicated situations, draw in other bodies using dashed lines – but still don't include any forces acting on these bodies (Figure 2.4). Otherwise you violate commandment 2. The dashed bodies remind us of how the interactions happened.

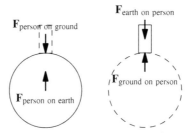

Figure 2.4 Step 3 (optional): Include a ghostly reminder of the forces' origins.

2.2 Practicing the Recipe

The freebody-diagram recipe will become more automatic after we make a couple more diagrams – of a falling stone (Section 2.2.1) and a bouncing ball (Section 2.2.2) – and fix a common but incorrect diagram of a person standing on the ground (Section 2.2.3).

2.2.1 Falling Stone

In the first example, a stone falls toward the ground.

▶ *Ignoring the atmosphere and air drag, what's its freebody diagram?*

You can draw this diagram without using the elaborate recipe of Section 2.1. However, by using the recipe anyway, you practice it in a situation where you already know the result and can therefore attend to the steps in the recipe.

Here, the stone is the primary body. Without any atmosphere, step 1 (handling the short-range interactions) could not get simpler: No body touches the stone, so it participates in no short-range, contact interaction.

In step 2 (handling the long-range interactions), the stone's only long-range interaction is the gravitational interaction with the earth. That identification therefore completes step 2a (finding the long-range interactions). The partner body, the earth, becomes the only secondary body.

In step 2b (drawing the long-range forces), we split the sole long-range interaction into its two sides: $\mathbf{F}_{\text{earth on stone}}$ and $\mathbf{F}_{\text{stone on earth}}$ (Figure 2.5). Each force, from (1.12), has magnitude mg, where m is the stone's mass. The force on the stone points from the stone to (the center of) the earth. Meanwhile, the force on the earth points from the (center of the) earth to the stone. Because the secondary body, the earth, interacts only with the primary body, the earth's diagram is also a freebody diagram.

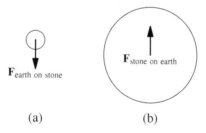

$\mathbf{F}_{\text{earth on stone}}$

$\mathbf{F}_{\text{stone on earth}}$

(a)   (b)

Figure 2.5 Freebody diagrams of the stone and the earth. (a) The stone (the primary body). (b) The earth (the secondary body).

These freebody diagrams are the simplest possible for a nonfree body: Each body experiences only one force, the unavoidable gravitational force. Such a body is said to be in free fall. Thus, the earth is in free fall too! And so is a thrown stone (with no air resistance) or a comet as it orbits the sun, even though neither's motion fits our everyday understanding of falling. A more explicit and less confusing description is *free gravitational motion*: motion free of any force except the gravitational force.

In the optional step 3, we remind ourselves of the reason for the interaction by drawing in the interacting bodies with dashed lines (Figure 2.6).

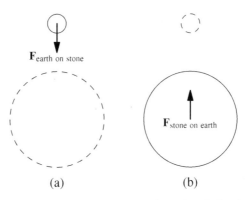

Figure 2.6 Freebody diagrams of the stone and the earth with ghostly reminders of the interaction partners. (a) The stone. The dashed earth reminds us of the stone's partner in the gravitational interaction. (b) The earth (with a dashed stone).

2.2.2 Bouncing Ball

The next example is a drag-free bouncing ball (Figure 2.7).

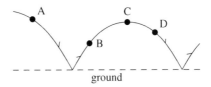

Figure 2.7 The path of a bouncing ball showing two bounces from the ground.

What are the freebody diagrams of the ball at locations A, B, C, and D?

At each location, the ball touches only the air, which we are ignoring. So, it is already a free body, and we skip all of step 1 (handling contact interactions).

Next come the long-range interactions (step 2). The ball's only long-range interaction (step 2a) is its gravitational interaction with the earth. This interaction's $F_{\text{earth on ball}}$ side goes on the ball's diagram. It has magnitude mg and points to the center of the earth. Thus, in answer to the triangle question: At each location A, B, C, or D, the ball's diagram gets a downward force mg – and no other force (Figure 2.8). The ball is in free gravitational motion (free fall).

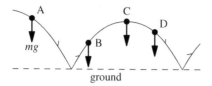

Figure 2.8 Freebody diagrams of the ball at each point. The only force acting on the ball is the gravitational force, mg downward.

You probably didn't need the recipe to draw these diagrams. But slowing down to follow the recipe anyway helps you avoid a common trap: including a bogus "force of motion" pointing in the direction of the ball's motion (try Problem 2.1).

The ball's freebody diagrams are, at least when the ball is in the air, identical to the falling stone's, yet the ball moves so differently from the stone. How? Why?

This question gets to the heart of Newtonian mechanics: to what made it new and to why we find it so difficult to grasp. Even though the sole force on these bodies is the same (mg downward), force merely *changes* motion – it doesn't cause motion directly. This counterintuitive idea, elaborated in Section 4.1 and *ad nauseam*, is worth befriending now.

Here, the ball's motion and the stone's motion *change* in the same way – but the stone started from rest, and the ball started with a sideways motion. The ball's sideways motion continues, not needing a sideways force to explain it. The constant sideways motion and changing downward motion – changing because of the downward gravitational force – together give the ball its curved path.

2.2.3 Standing on the Ground, Diagrammed Badly

Figure 2.9 shows a common incorrect freebody diagram.

What's wrong with the diagram (other than the lack of labels)?

The person is not free and is still touching the earth! Thus, step 1a got skipped. That procedural error leaves the force arrows at the contact point ambiguous. Does the lower downward force inside the box (the person) act on the earth or on the person? It acts on the earth, as you saw in Section 2.1, but the flawed diagram hides this important information. Similarly, does the upper upward force inside the circle (the earth) act on the earth or on the person?

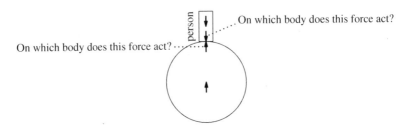

Figure 2.9 A common (but incorrect) freebody diagram.

This example is meant to remind you of step 1a: To make a freebody diagram, tear the primary body from all bodies that touch it by replacing each contact with a contact interaction. Then, in step 1b, place on the primary diagram exactly one side of each contact interaction.

2.3 A Subtle Puzzle: Bumblebees in a Box

As this chapter's final example of making a freebody diagram, we begin to solve the following puzzle. It concerns a truck carrying among its cargo a box of bumblebees. (No bees were harmed in the dreaming up of this experiment: The box has holes so that the bees get plenty of oxygen.) The truck gets pulled over to the weighing station to check whether it's too heavy. The road's limit is 10 metric tons (10^4 kilograms or 10^7 grams). Alas, when put on the scale, the truck weighs 10.00001 tons, which is 10 grams over the limit.

The puzzle: Can the driver reduce the truck's weight with the help of the bees? For example, what if the driver whacks the box when the bees are sleeping peacefully on the floor of the box, waking up the bees who then take off? While the bees take off, is the truck's weight lower?

Solving this puzzle requires, among other ideas, understanding weight and the distinction between it and the gravitational force. Thus, we'll solve it fully (in Section 7.4.5) after having discussed those ideas. However, the first step in the solution is a freebody diagram – which we can make now.

In making a freebody diagram, a key step is choosing the correct primary body. It might be the truck, the weighing scale, the box, or one of the bees.

It might even combine these bodies. For nothing in Newton's laws requires that a primary (or secondary) body be a single solid body. In astronomy, a useful primary body is the billions of stars in our galaxy; here, it might be all the bees as they fly around (with a mental box imagined around them). Or it could be all

the bees along with the physical box surrounding them. In short, the primary
(or any secondary) body can be composite.

To winnow the possibilities for the primary body's composition, let's at least
eliminate the truck. The box of bees now sits directly on the weighing scale (think
an old-fashioned bathroom scale; see Figure 2.10), which displays 10.01 kilograms (10 grams too high). The revised puzzle presents the same conundrum as
does the original puzzle: Can banging on the box, or another action that affects
the bees' motion, reduce the weight below 10 kilograms?

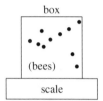

Figure 2.10 The box with its bees sitting on the weighing scale. When the bees
are resting on the bottom of the box, the scale displays 10.01 kilograms.

As a rule of thumb, use the freedom in choosing the primary body to winnow
this body's interactions – as long as the remaining interactions include all forces
of interest. But what are those forces here? As you learn in Section 7.4, a (spring)
scale measures the normal force **N** on its top surface and displays this force's
magnitude, N (after converting N to mass units using N/g). Thus, the interesting
force is the normal force on the scale's top surface.

Then the surprising best choice for the primary body is the box along with its
bees. This composite body has only two interactions (Figure 2.11) – a pleasantly
small number. Its short-range, contact interaction is with the scale. Its long-range,
gravitational interaction is with the earth.

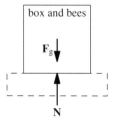

Figure 2.11 Freebody diagram of the box and its bees. The ghostly outline reminds
us of the origin of the normal force (the scale).

Thus, its freebody diagram, like the freebody diagram of a person standing on
the earth (Section 2.1), has only two forces: the normal force *due to* the scale
and the gravitational force due to the earth. This normal force's magnitude, N, is

also the magnitude of the normal force *on* the scale (a consequence of the third law). Thus, the primary body's interaction does include the force of interest.

The puzzle then becomes the following: Can N be reduced by getting the bees to change their motion? Our carefully chosen freebody diagram, though not a solution by itself, has helped us to translate the puzzle into Newtonian language and will make the puzzle solvable – a promise redeemed in Section 7.4.5.

2.4 Problems

2.1   A common mistake in drawing the forces on the bouncing ball of Figure 2.7 is to include a force in the direction of motion – for example, downward and to the right at location A or directly to the right at location C. Based on what you know about forces and interactions from Chapter 1, how would you convince a wayward student not to include such a force?

2.2   A roller-coaster car, zooming around a loop-the-loop track, is right now at the top of the loop (Figure 2.12a). Ignoring air resistance, its freebody diagram (Figure 2.12b) has, surprisingly, only downward forces. (In Chapter 7 and Problem 7.19, you learn how a car can stay on the track without any upward force.) Make labeled freebody diagrams for the track and the earth (which includes the ground).

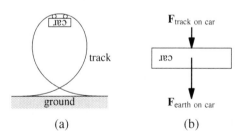

(a)   (b)

Figure 2.12 A roller-coaster car and its freebody diagram. (a) The car zooming around a loop-the-loop track. (b) Freebody diagram of the roller-coaster car when it's at the top of the loop.

3

Newton's First Law: Permission to Use Newton's Second Law

Now you know that every force is one side of an interaction (Newton's third law). The next, surprising, step is to imagine a body with no interactions. Making that mental picture is the first step in using Newton's first law.

> **Newton's First Law.** A body on which no force acts continues to move at constant velocity (in a straight line at constant speed).

The first law provides an essential test. If our reference frame – our three-dimensional space – passes this test, Newton's second law is valid in that reference frame, and we can freely use the second law. If our reference frame fails the test, we cannot use the second law. The second law is the workhorse of Newtonian mechanics. It describes mathematically what forces do (namely, their effect on motion). To use Newtonian mechanics for understanding and planning motion, we have to know whether we can use the second law.

To that end, you first learn what a reference frame is (Section 3.1). Second, you learn how to test a reference frame using the first law (Section 3.2). Third, you meet common failing reference frames (Section 3.3). Finally, you learn how to make new passing reference frames (Section 3.4).

3.1 Reference Frames

A reference frame is a system of viewing the world. In Newtonian mechanics, it's an infinite three-dimensional Euclidean space free to move (to change its position) and rotate (to change its orientation). Think of it as an infinite three-dimensional lattice of rigid metal rods. Then the "reference frame of (a rigid) object X" means the lattice attached to X and moving and rotating with X.

The lattice's motion and rotation can be described by choosing a node called the origin (where the object X is) and choosing three mutually perpendicular rods growing out of that node (Figure 3.1). Paint the first, say, red, the second green, and the third blue. Then you can describe the lattice's motion by giving the location of the origin and the orientation of the red, green, and blue rods.

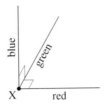

Figure 3.1 Describing the location and orientation of a reference frame. The origin is X. The red, blue, and green rods are mutually perpendicular, with the green rod going into the page. The location of X and the orientation of the three rods specify the location and orientation of the reference frame.

However, nothing is special about this node or the three painted rods. In particular, no observer sits at the origin. Rather, in Newtonian mechanics, the observer is omnipresent (is everywhere) and sees the entire infinite space at a glance. That said, in deciding how a body moves in a reference frame – the essential step in the first-law test – I do mentally place an observer at the origin and orient the observer according to the frame's orientation. Often, I make myself the observer – maybe because I am self-absorbed but also because doing so strengthens my feel for the frame's motion and rotation. Here are examples of reference frames.

1. *The frame of the earth.* This frame moves with the earth around the sun and rotates with the earth as the earth spins on its axis. This frame, the most common in everyday life, is usually intended when no frame is specified.

2. *The frame of a train car moving at constant speed on a straight track.* This frame moves at constant velocity and, because the track is straight, does not change its orientation. But I just smuggled in a subtle point. The descriptions "constant velocity" and "does not change its orientation" depend on who's looking. More formally, these descriptions depend on the base reference used to describe the train car's motion and rotation. If unspecified, the base frame is usually the frame of the earth (frame 1), as it is here.

3. *The frame of an elevator falling freely toward the earth.* This frame starred in the development of Einstein's theory of gravitation, general relativity. Einstein used it to explain how gravity isn't a force at all! This fascinating story is touched upon in Section 8.3.2.

4. *The frame moving with the center of the earth but not rotating with the earth.* This frame (Figure 3.2), unlike frame 1, does not change its orientation. Again, whether it changes its orientation depends on who's looking, on the base reference frame. Here, that frame is the sun's frame. Then "not rotating with the earth" means that, looking down onto the solar system from above, the lattice of rods moves but does not spin: The red rod ("r"), for example, always points to the right.

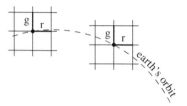

Figure 3.2 A nonrotating lattice whose origin is moving. The green and red rods, as judged in the base reference frame, maintain their orientation.

5. *The frame fixed to a rotating merry-go-round (a roundabout).* This frame, whose origin stays at the center of the merry-go-round, changes its orientation as it rotates with the merry-go-round.

3.2 Applying the Test

For any frame, including the common frames in Section 3.1, you need to know whether Newton's second law is valid. That is, does the second law connect the forces on a body to the body's change in motion? This question is settled by the first-law test. This test has two steps.

1. *Prepare an isolated body.* You choose a body, the test body, and remove from it all interactions. This removal has two parts, corresponding to short-range and long-range forces (Section 1.2.3). To remove the short-range interactions, ensure that the test body does not touch any other bodies. To remove the long-range interactions, ensure that no other body pulls it gravitationally (I am ignoring long-range electromagnetic interactions). The isolated test body's freebody diagram is now simple: It's just the test body without force arrows.

2. *Observe the body's motion.* In the candidate reference frame, observe how the test body moves. If it moves at constant velocity – at constant speed and in a straight line – then the reference frame passes the test. It's an inertial reference frame. Only in an inertial frame is Newton's second law valid.

In testing a reference frame, it's important to distinguish necessary from sufficient conditions. It's sufficient to prepare an isolated body (a body with no interactions) and observe its motion. However, and surprisingly, it's not necessary *to find* an isolated body: Even in a world without isolated bodies, the reference frame used to describe that world might still be an inertial reference frame, and the second law might still be valid. But how would you then know? You perform the first-law test in your imagination, as a thought experiment rather than a physical experiment.

This issue isn't merely theoretical. The reference frame of greatest interest, the frame of the earth (frame 1 in the example frames of Section 3.1), has no isolated bodies: Each of its bodies participates in a gravitational interaction with the earth. How then can you test whether the second law is valid?

The most thorough answer is to use a test body far, far away from the earth – say, halfway to Mars – where the earth's gravity has only a tiny effect. This answer, a physical experiment, is expensive.

Instead, use a thought experiment. Consider only a part of the problem by limiting the reference frame to a small patch of the earth's surface and ignoring vertical motion. Because the gravitational force pulls the body only vertically, it creates motion outside the scope of this two-dimensional reference frame. Similarly, the normal force also acts outside the scope of this reference frame.

Then isolate the test body only in the horizontal directions (the directions within the reference frame). Ignoring gravity leaves no long-range interactions. To remove the nonignored portions of the short-range interactions, namely their horizontal portions (including friction and air resistance), place the body on a sheet of perfectly smooth ice in a world with no atmosphere. Thus, imagine the body on an perfectly smooth and iced-over lake on an airless earth.

Now apply the test (step 2). Perhaps after hitting the test body to get it moving (although you could also leave it alone and still perform the test), see whether the body moves at constant velocity – at constant speed in a straight line. If you are attached to the earth's surface (for example, standing on the icy lake) and extrapolate from everyday experience of smooth surfaces to ideal, perfectly smooth surfaces, you see the body slide across the ice at constant velocity. Thus, the two-dimensional reference frame of a small patch of the earth's surface passes the test and is an inertial reference frame.

(To be extremely careful, we need one more condition: that the test last not too long, at most a small fraction of a day, so that the test body's path isn't appreciably affected by the earth and thus the ice sheet rotating underneath. Another way of expressing this condition is that the Coriolis force, discussed in Section 8.1, not appreciably affect the test body's velocity.)

But what about the third, vertical dimension? Is the three-dimensional frame still inertial?

The third dimension is indeed trickier because of the vertical forces that were just ignored in the two-dimensional analysis. As a test body, take an apple about to fall from a tree. Just after the twig holding up the apple breaks, only one force acts on the apple: gravity (in an airless world without air resistance). To isolate this test body, perform the thought experiment of turning off gravity. The apple then hangs in midair (if I can describe it that way in a world without air). "No motion" is a special case of constant-velocity motion. Thus, the three-dimensional reference frame is inertial.

The earth rotates on its axis, so how can it be an inertial reference frame?

It isn't, but it's close. For example, even when I throw a rock with all my (admittedly limited) might, I can hardly detect in its trajectory the effect of the earth's rotation. However, for travel over longer distances – for example, for spacecraft, long-range artillery, or migrating birds – the effect is larger and necessary to include. It's touched upon in Section 8.1. Until then, I'll treat the frame of the earth as an inertial reference frame.

3.3 Noninertial Frames

Not every reference frame qualifies as an inertial reference frame – otherwise the first law would be pointless. To make a noninertial reference frame, stand at the center of a rotating merry-go-round, and consider motion in just the two horizontal dimensions. A rock – the test body – lies on perfectly smooth ice just beyond the merry-go-round. It has no horizontal interactions, so step 1 is done. Before doing step 2, note that a person standing on the ice next to the rock sees the rock move at a constant (zero) velocity – reconfirming that this person's frame, which is the earth's frame, is an inertial frame.

Now for step 2: You, at the center of and rotating with the merry-go-round, see the same rock move in a circle centered on you. If the merry-go-round rotates at a constant rate, you will see the rock move at a constant speed. However, constant speed is insufficient for passing the first-law test. The test requires constant *velocity*: constant speed and fixed direction. Here, the rock's direction changes as it moves around the circle. Thus, the reference frame of the merry-go-round, because of its rotation, fails the test and isn't an inertial reference frame. In that reference frame, the second law cannot be used.

(Inverting the finicky condition about the ice-sheet frame in Section 3.2, that the test not last too long lest the earth's rotation affect the test body's path appreciably: Here the test should last at least a significant fraction of the merry-go-round's rotation period. Then the reference frame's rotation will affect the test body's path appreciably.)

As another example of where the second law shouldn't be used, imagine sitting in that vehicle rounding a turn (Section 1.2.1). Its analysis invites frequent confusions – for example, spurious forces that could have been avoided by applying the first-law test first.

▷ *Sitting in the vehicle, are you in an inertial reference frame?*

To decide, you need a test body – for example, the isolated rock sitting on an icy lake. From your point of view as you move and rotate with the vehicle, the rock moves in a curved path. Thus, you are not in an inertial reference frame. Therefore, Newton's second law is invalid in this frame. This information is useful because dangerous misconceptions, particularly about centrifugal and centripetal forces, are fostered by using the second law anyway (more details are in Section 7.3.2). Using the first-law test nips these problems in the bud.

3.4 Making New Inertial Frames

Finding an inertial reference frame is the precondition for using Newton's second law, the workhorse of the three laws. However, you have so far met only one inertial reference frame, the frame fixed to the earth – which is anyway only an approximate inertial frame. If Newton's laws applied only in that one frame, they would hardly be worth calling laws.

◁ *Are there other inertial frames?*

Yes! Any reference frame moving at constant velocity (and not rotating) with respect to a base inertial frame is also an inertial frame. A classic example of such a derived inertial frame is a train moving at constant velocity (Figure 3.3). "Constant velocity" means that the track is perfectly straight and smooth: Any turn or bump would change the train's direction, even if slightly.

To test this frame using the first law, imagine a huge icy lake – the base inertial frame – with an ideal straight train track running across it. The test body is a rock moving north at 30 meters per second (108 kilometers, or roughly 65 miles, per hour). The train moves east at 40 meters per second (144 kilometers, or roughly 90 miles, per hour).

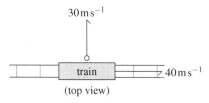

(top view)

Figure 3.3 Train traveling east past a rock moving north. The train's velocity and the rock's velocity are measured relative to the base inertial frame, the icy lake across which the track runs.

In the reference frame of the train, what's the rock's velocity?

If the rock were fixed to the ice (instead of moving north), a train passenger would see the rock move backward (west) at 40 meters per second. Because the rock is actually moving north at 30 meters per second (relative to the icy lake), the train passenger sees the rock move with two motions combined (Figure 3.4):

$$\mathbf{v} = 30 \text{ m s}^{-1} \text{ north} + 40 \text{ m s}^{-1} \text{ west.} \tag{3.1}$$

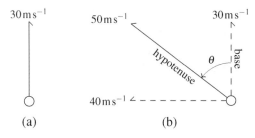

(a)　　　　　　　　　(b)

Figure 3.4 The rock's velocity as measured in the two inertial frames. (a) In the base frame (the icy lake), the rock moves due north. (b) In the train frame, the rock moves west as well, so its velocity makes a nonzero angle θ past due north.

Alternatively, this velocity, like any vector, can be described by its magnitude (the speed) and its direction. From the Pythagorean theorem, the rock's speed v (as seen in the train frame) is 50 meters per second:

$$v = \sqrt{(40 \text{ m s}^{-1})^2 + (30 \text{ m s}^{-1})^2} = 50 \text{ m s}^{-1}. \tag{3.2}$$

(The vector sum of the two velocities makes a magnified 3–4–5 right triangle.)
The rock's direction is given by the angle θ. From the definition of cosine,

$$\cos \theta \equiv \frac{\text{base}}{\text{hypotenuse}} = \frac{30 \text{ m s}^{-1}}{50 \text{ m s}^{-1}} = 0.6. \tag{3.3}$$

Thus, θ is approximately 55 degrees:

$$\theta = \arccos 0.6 \approx 55°. \tag{3.4}$$

So, as a second answer to the triangle question: The rock moves at 50 meters per second nearly northwest (more precisely, at a compass heading of approximately 305 degrees).

Because the rock's speed, given by (3.2), and direction, given by (3.4), remain fixed, the train frame passes the first-law test.

This calculation illustrates how to calculate a body's velocity in a new nonrotating reference frame. The general recipe is

$$
\begin{pmatrix} \text{velocity of} \\ \text{a test body} \\ \text{as seen in the} \\ \text{new frame} \end{pmatrix} = \begin{pmatrix} \text{velocity of} \\ \text{the test body} \\ \text{as seen in the} \\ \text{base frame} \end{pmatrix} - \begin{pmatrix} \text{velocity of} \\ \text{the new frame} \\ \text{as seen in the} \\ \text{base frame} \end{pmatrix}. \quad (3.5)
$$

When the velocity of the new frame (as seen in the base frame) is constant, the new frame gives the same first-law test result as the base frame gives. A test body that moves at constant velocity in the base frame will move at constant velocity in the new frame. Similarly, a test body that moves at nonconstant velocity in the base frame will move at nonconstant velocity in the new frame. Thus, the new frame is an inertial reference frame if and only if the base frame is.

In summary, to make a new inertial reference frame, make a reference frame moving at constant velocity relative to a known inertial frame (and not rotating). Now that you know several inertial reference frames and how to make others, you have places to use the second law – the workhorse of mechanics and introduced in the next chapter.

3.5 Problems

3.1 At a few positions in the earth's orbit, draw the rod lattice for frame 1 of Section 3.1: the frame that moves with the earth around the sun and rotates with the earth as the earth spins on its axis. How do the lattices compare to the lattices for the frame moving with the center of the earth but *not* rotating with the earth (frame 4 of Section 3.1)?

3.2 If the train accelerates forward, moving ever faster than 40 meters per second, what does the rock's path look like roughly (as seen in the train frame)? Is this accelerating-train frame an inertial reference frame?

4

Introducing Newton's Second Law

You are here, ready to use Newton's second law, because your reference frame passed the first-law test (Section 3.2): An isolated body, one with no forces acting on it, moved in a straight line at constant speed. However, what happens to its motion when forces do act? That question, the most important in mechanics, is answered by Newton's second law.

This law contains three crucial ideas. First, force *changes* a body's motion (Section 4.1). Second, this change is produced by the *sum* of the forces on the body (Section 4.2). Third, the change is inversely proportional to the body's *mass* (Section 4.3). As I elaborate each idea, I will translate it into symbols. Thus, by the end of this short chapter, you'll know the formal statement of Newton's second law and the meaning of its every symbol.

4.1 Force Changes Motion

The second law's first idea is that force *changes* motion. Motion here means velocity (a vector). Thus, force changes velocity.

Unchanging or constant-velocity motion requires no explanation – or, at least, finds none in Newton's laws. For you can give a body any velocity that you want merely by changing to a new inertial reference frame (Section 3.4), and the second law is equally valid there. However, the *change* in a body's velocity is the same in any inertial frame and therefore worth explaining. The second law provides the explanation and names the cause: force.

Velocity **v** is a vector. The rate at which velocity changes is also a vector. This vector is the acceleration **a**. It's discussed fully in Chapter 6. For now, keep it in mind simply as measuring how velocity changes. Then remember:

47

Newton's Second Law (Idea 1). Force causes acceleration (not velocity).

Because velocity is a vector, having magnitude and direction, it can change in two ways. First, its magnitude, the speed, can change. When a body speeds up or slows down, it has a nonzero acceleration. Second, the direction of motion can change: Imagine a model train moving at constant speed around a circular track. The train, even though its speed is fixed, has a nonzero acceleration because it changes its direction of motion. This second way is often overlooked because, in everyday language, *accelerating* and *decelerating* refer only to changing speed. But the physics concept of acceleration, which the second law says is caused by force, also includes changing direction.

The Newtonian idea, that force causes acceleration, ranks first among the three ideas because it's the most misunderstood. The misunderstanding has ancient roots that embed it deep into our psyche.

Once upon a time, millions of years ago, there was a tribe of hominids on the African savanna, each hominid leaning idly against a favorite tree (Figure 4.1). The tribe was divided on the right view of force. One faction believed that force causes velocity. The other faction – Newtonian thinkers eons before Newton – believed that force changes velocity (that force causes acceleration).

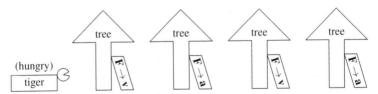

Figure 4.1 Hominids on the savanna leaning idly against the trees. They differ on whether force causes velocity or acceleration. When the tiger appears, the **F** → **v** crowd runs hard and keeps on running. The **F** → **a** crowd pushes hard but only once against the tree; these prematurely Newtonian thinkers had few, if any, children.

Along came a hungry tiger seeking a tasty dinner. The force-causes-velocity faction thought, "To get away, I need a large speed, so I had better apply a large force," and ran hard from the tiger. The innovative, Newtonian faction, believing that force is needed only to change velocity, pushed hard against their trees just once, relying on the resulting speed to carry them to safety. The tiger made short work of this belief and its adherents, finding among these innovators its dinner.

The Newtonian thinkers had few, if any, children and, after many generations, went extinct. Our ancestors were all force-causes-velocity thinkers – making Newtonian mechanics hard for us to learn today, even though tigers no longer eat the ones who learn it.

4.2 What Matters Is Net Force

Force causes acceleration; or, more explicitly, the force on a body causes the body's acceleration. But this statement points to a contrast. Each body has, at any moment, a single acceleration. However, few bodies participate in only one interaction and experience only one force. What happens when a body participates in several interactions and experiences several forces? This question is answered by the second crucial idea in Newton's second law:

> **Newton's Second Law (Idea 2).** The effect of several forces is the effect of their vector sum.

This vector sum is called the net force. Here, "net" takes its commercial meaning of "remaining after a deduction, such as a tax or discount, has been made" [21, p. 1178] – for example, "net weight" (weight remaining after deducting the packaging's weight). To transfer the idea to mechanics, generalize "deduction" to mean "addition, but keeping track of signs." For forces, which as vectors have no sign, the idea of sign generalizes to direction. Then "adding, but keeping careful track of direction" exactly describes how to compute their vector sum. Another name for this vector sum, common in Commonwealth countries, is "resultant force." Despite being a Commonwealth citizen, I prefer "net force" because it tells me *how* to compute the result: by (vector) addition.

In symbols, the first two ideas say

$$\sum_i \mathbf{F}_i \text{ causes } \mathbf{a}. \tag{4.1}$$

The $\sum$ symbol, a Greek uppercase sigma, means summation; its subscript i ranges over 1, 2, 3, ..., one number for each force acting on the body; and $\mathbf{F}_i$ is the ith force. Thus, $\sum_i \mathbf{F}_i$ is shorthand:

$$\sum_i \mathbf{F}_i \equiv \mathbf{F}_1 + \mathbf{F}_2 + \cdots. \tag{4.2}$$

For example, when you stand on the ground (Section 1.1), the two forces on you – the gravitational and the contact force – have identical magnitudes and opposite directions (as you'll learn in Section 5.1), so their vector sum is zero:

$$\sum_i \mathbf{F}_i = \mathbf{F}_{\text{earth on you}} + \mathbf{F}_{\text{ground on you}} = 0. \tag{4.3}$$

A common abbreviation for the vector sum $\sum_i \mathbf{F}_i$ is $\mathbf{F}_{\text{net}}$. With that notation, the first two ideas in Newton's second law have a compact form:

$$\mathbf{F}_{\text{net}} \text{ causes } \mathbf{a}. \tag{4.4}$$

When you see acceleration – a body change speed or direction – remember that several forces could act on the body and that their *sum* causes the acceleration.

4.3 More Mass Means Less Acceleration

That net force causes acceleration cannot be the whole story of the second law, of the effect of force on acceleration. A 10-newton force on a train car, even one sitting on a smooth track, hardly changes the train car's motion. The same force on a book, even on a doorstopper like *War and Peace* in large print, has a much greater effect. This intuition is formalized in the third crucial idea.

> **Newton's Second Law (Idea 3).** The effect of the net force on a body is inversely proportional to the body's mass.

The three ideas combined together are Newton's second law. In symbols, it says

$$\frac{1}{m} \sum_i \mathbf{F}_i \rightarrow \mathbf{a}, \tag{4.5}$$

where m is the body's mass. The $\mathbf{F}_{net}$ shorthand gives it a more compact form:

$$\frac{\mathbf{F}_{net}}{m} \rightarrow \mathbf{a}. \tag{4.6}$$

In (4.5) and (4.6), our intuition about mass lives in the factor of $1/m$. Thanks to the precision provided by this factor, the slightly ambiguous word "causes" could be replaced with an arrow ($\rightarrow$). The *causal arrow* contains two ideas.

1. *The left side causes the right side – and not vice versa.* An equals sign ($=$) would permit either or no direction of causation. But the second law specifies that the direction of causation is from force to acceleration (Section 4.1).

2. *The expressions on the two sides of the arrow are equal in value.* A profound consequence of this equality is that, because the left and right sides of the arrow are vectors, equality means equality in magnitude and in direction. Thus, and this point is easily forgotten (particularly when a body moves in a circle), the net force and the acceleration point in the same direction. So, if you know the one quantity's direction, you know the other's.

   The equality invoked by the arrow refers to the two sides' values rather than to their natures. Their natures are necessarily different because one side causes the other. For example, the net force on a falling 5-kilogram doorstopper of a book might be 45 newtons downward: 50 newtons downward due to the gravitational force and 5 newtons upward due to air drag. Then the second law says that

$$\frac{45 \, \text{N downward}}{5 \, \text{kg}} \rightarrow 9 \, \text{m s}^{-2} \, \text{downward}. \tag{4.7}$$

Because a newton per kilogram is a meter per second squared, the left side has the same value as the right side (but their natures are different).

To summarize the three ideas in symbols, each idea is highlighted in black type.

Force changes motion (causes acceleration): $\dfrac{1}{m}\sum_i \mathbf{F}_i \to \mathbf{a}$.

What matters is net force: $\dfrac{1}{m}\sum_i \mathbf{F}_i \to \mathbf{a}$.

More mass means less acceleration: $\dfrac{1}{m}\sum_i \mathbf{F}_i \to \mathbf{a}$.

Combining these three symbolic representations gives the full statement of the second law (4.5) or, using the $\mathbf{F}_{net}$ notation, (4.6).

4.4 Alternative Forms of the Second Law

The second law is often written a similar but less informative form:

$$\sum \mathbf{F}_i = m\mathbf{a}. \tag{4.8}$$

Compared to my preferred form (4.5), here the mass and acceleration are combined on the right side, and the causal relation is replaced by an equality. This form leaves ambiguous why the two sides are equal. Are they equal because the body's acceleration causes the net force acting on the body (in proportion to the body's mass)? Or are they equal because the net force causes the acceleration (in inverse proportion to mass)? The causal-arrow version (4.5) specifies the second option: force causes acceleration.

Why is the equality form (4.8) so common?

As pointed out by Judea Pearl [16, p. 4], a pioneer in the probabilistic analysis of causation, mathematics has no standard notation for causation. Therefore, when we express physical laws in mathematics, in honor of Galileo's maxim that the book of nature is written in mathematics, causation easily finds itself homeless.

An even simpler form of the second law, which you won't find in (correct) textbooks, is

$$\mathbf{F} = m\mathbf{a}. \tag{4.9}$$

Phonetically: "Eff EE-kwuhlz emm ay." This form is how many students and teachers, including me, mentally hear the second law. It shares the $\sum \mathbf{F}_i = m\mathbf{a}$ form's vagueness about causation; it also does not specify which force to use in the left-hand side as the $\mathbf{F}$. Thus, it's easy to forget that the correct "force" is the net force – the sum of all forces on a body.

For these many reasons, my recommended form is (4.5). It contains two new large ideas: summation of forces and acceleration. Summation of forces is elaborated in Chapter 5, where, in order to focus just on that idea, we study motion without acceleration. The bodies therefore move at constant velocity – in a straight line at constant speed. (The constant speed could be zero, so the restriction to zero acceleration contains the common case of a body just sitting there.) In Chapter 6, we study the second large idea, acceleration. Then, in Chapter 7, we combine summation of forces with acceleration in order to use the full second law.

5

Newton's Second Law with Zero Acceleration

As you learned in Chapter 4, Newton's second law (4.5) connects the sum of the forces on a body to the body's acceleration. In this chapter, in order to concentrate on the summation of forces, we apply the second law only to nonaccelerating bodies. In that case, the second law specializes to

$$\frac{1}{m} \sum_i \mathbf{F}_i \to \mathbf{a} = 0 \quad \text{or} \quad \mathbf{F}_{\text{net}} = 0. \tag{5.1}$$

A body's acceleration can be zero for two qualitatively different reasons. First, no force acts on the body – the situation that we construct in order to apply the first-law test and decide whether the second law holds (Section 3.2). But we already know that the second law holds: We would never even use Newton's second law without having applied the first-law test! Second, forces do act but are arranged just right to prevent the body from accelerating. In other words, the net force is zero. This more interesting possibility – called statics – is the subject of this chapter.

5.1 Standing on Level Ground

As an example illustrating a common pitfall and how to avoid it, imagine yourself once again standing on the ground. Because your velocity is constant (in particular, it's zero), your acceleration is zero. Thus, this situation is within the scope of our restricted second law (5.1).

The first step in applying the second law is a freebody diagram (Figure 5.1). This diagram appeared in Section 2.1, where you learned how to make freebody diagrams. Back then, the contact force's magnitude was unknown. However, you might have guessed that it's *mg*. If so, you are right.

53

$$F_{\text{contact}}$$

Figure 5.1 Freebody diagram of you standing on the ground. Because you participate in two interactions, you experience two forces.

➤ *But why? Is it because of Newton's third law about equal and opposite forces?*

No! Newton's third law relates the two sides (the two forces) of *one* interaction.
It relates, for example, the gravitational force of the earth on you (mg downward) to the gravitational force of you on the earth (mg upward). Those two
forces are indeed equal in magnitude and opposite in direction. However, the
gravitational force on you and the contact force on you belong to different interactions. Although these two forces are equal in magnitude and opposite in
direction, this circumstance cannot be because of the third law.

The correct justification uses the second law: The acceleration is zero, so the
net force must be zero. That justification is a logical deduction (it lives in our
mental world). The causality (the physical world) operates in the other direction:
Zero net force produces zero acceleration. Knowing the acceleration, you use
knowledge of the causality to deduce that the net force is zero:

$$\mathbf{F}_g + \mathbf{F}_{\text{contact}} = 0. \tag{5.2}$$

Solving for $\mathbf{F}_{\text{contact}}$,

$$\mathbf{F}_{\text{contact}} = -\mathbf{F}_g. \tag{5.3}$$

Because $\mathbf{F}_g$ equals mg downward, $\mathbf{F}_{\text{contact}}$ equals mg upward. You knew this
already, but – and in answer to the triangle question – now you also know that
it's a (logical) consequence not of the third law but rather of the second law and
of the condition of zero acceleration. This distinction is important because it
specifies the conditions under which the conclusion (5.3) holds. The conclusion
fails when your acceleration is nonzero – as you will learn in Section 7.1.4 and
(for a large surprise) Section 7.1.7.

The contact force on you, as you found in Problem 1.2, is a passive force:
It adjusts itself in response to the known gravitational force. We deduce its
magnitude mentally. The ground, through the adjustment process, deduces its
magnitude physically and comes to the same conclusion as we do: mg upward.
(How the ground makes this physical deduction is discussed in Section 7.1.6.)

This analysis and, indeed, all the analyses in this chapter illustrate one ideal type in using the second law: You know everything about a body's acceleration. From that knowledge, you use the second law in reverse to infer the forces acting on the body. Thus, I call this analysis direction type I (inferring).

In the opposite ideal type, you know everything about the forces acting on a body, and you use the second law in the forward direction to calculate the body's acceleration (how its velocity changes). Thus, I call this analysis type C (calculating). Its first example will be the freely falling stone of Section 7.1.1.

Most uses of the second law mix these ideal types. You know something, but not everything, about a body's acceleration and know something, but not everything, about the forces acting on the body. Using the second law, you infer and calculate the missing information about the forces and acceleration, ensuring that the complete set is consistent. Examples will include sliding on level ground with friction (Section 7.1.2) and sliding down a frictionless slope (Section 7.1.4). But let's return to the world of zero net force.

5.2 Standing on a Hill

If you stand on a hill (Figure 5.2a) rather than on level ground (a hill with $\theta = 0$), as in Section 5.1, your freebody diagram remains the same! Your acceleration hasn't changed (it's still zero). Nor has the gravitational force. To balance the gravitational force, the passive contact force still adjusts itself to be mg upward.

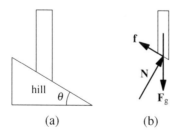

(a)   (b)

Figure 5.2 Standing on a hill. (a) You stand upright on the hill, which has inclination angle θ relative to horizontal. (b) Your freebody diagram is the same as your freebody diagram standing on level ground (Figure 5.1). The vertical contact force can be broken into two perpendicular portions: one perpendicular to the hill, called the normal force $\mathbf{N}$, and the other parallel to the hill, called static friction $\mathbf{f}$.

On the hill, however, the contact force has a useful reinterpretation based on breaking it into two portions (Figure 5.2b): a portion perpendicular to the hill and a portion parallel to the hill. The perpendicular portion is called the normal force – using, as mentioned in Section 1.3.4, the mostly archaic meaning of "normal" as "perpendicular." The parallel portion is called static friction. But these forces are really just portions of a single contact force.

You can also break up the gravitational force into portions perpendicular and parallel to the hill (Figure 5.3a). Its perpendicular portion balances the normal force. Its parallel portion balances static friction. Because the perpendicular portion has magnitude $mg \cos \theta$ (see Problem 5.1), so does the normal force:

$$N = mg \cos \theta. \tag{5.4}$$

Similarly, because the parallel portion has magnitude $mg \sin \theta$ (what you also show in Problem 5.1), so does the static-friction force:

$$f = mg \sin \theta. \tag{5.5}$$

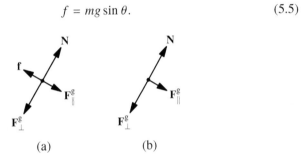

(a)   (b)

Figure 5.3 Splitting the gravitational force as well into perpendicular and parallel portions. (a) With the hill rough enough: Gravity's perpendicular portion $\mathbf{F}_\perp^g$ balances the normal force $\mathbf{N}$. Meanwhile, its parallel portion $\mathbf{F}_\parallel^g$ balances static friction $\mathbf{f}$. (b) After oiling the hill: $\mathbf{F}_\perp^g$ still balances $\mathbf{N}$. However, $\mathbf{f}$ vanishes, leaving a nonzero net force that accelerates you down the plane (Section 7.1.4).

Now you know what happens to you, in Newtonian terms, when the hill becomes frictionless (Figure 5.3b). Static friction $\mathbf{f}$ vanishes, leaving the parallel portion $\mathbf{F}_\parallel^g$ unbalanced and free to accelerate you downhill. Because this situation involves acceleration, it's discussed further in Chapter 7 (in Section 7.1.4).

5.3 Standing in a Steadily Descending Elevator

Now imagine standing in an elevator descending in a straight line at constant speed (thus, moving at constant velocity).

▶ *Is the magnitude of the contact force on you greater than, equal to, or less than mg (its magnitude when you stand on the ground, as you found in Section 5.1)?*

Two analyses are possible and agree that the magnitude equals mg. First, you might notice that, even though you are moving, your acceleration is zero. Therefore, the two forces acting on you, the gravitational force and the contact force, balance; so, the contact force has magnitude mg.

Second, you might notice that the elevator is an inertial reference frame: It moves at constant velocity with respect to the earth frame, an (approximate) inertial reference frame (as discussed in Section 3.1). In the elevator frame, you don't move at all. So, you might as well be standing on the ground, where the contact force, as we determined in Section 5.1, has magnitude mg.

The steadily descending elevator illustrates again the irrelevance to Newton's laws of velocity. Although your velocity standing in the elevator differs from your velocity standing on the ground (measuring both velocities in the earth frame), this difference implies nothing about the net force on you. In contrast, acceleration does matter: Because your acceleration is the same in both situations (zero), the net force on you must also be the same (zero).

5.4 Bicycling on Level Ground

In the preceding examples, you experienced only two forces. To practice with three forces, imagine bicycling on level ground at constant velocity. Give yourself a fast speed: 15 meters per second (54 kilometers, or about 33 miles, per hour) – easier to imagine than to achieve!

▶ *What forces act on the composite body of you plus the bicycle?*

Any such forces belong on a freebody diagram. The procedure for making one, slightly abbreviated from Section 2.1, is as follows.

1. Tear the body from all touching bodies, mentally replacing each contact with a contact interaction. Draw once force for each contact interaction.

2. Find the body's long-range interactions. Draw one force for each.

Here, the composite body touches two other bodies: the ground and the air. (Bicycling at 15 meters per second will convince anyone that the air matters: Overcoming drag demands a huge effort.) Thus, the body participates in two contact interactions and experiences two contact forces: $\mathbf{F}_{ground}$ (short for $\mathbf{F}_{ground\,on\,body}$) and $\mathbf{F}_{drag}$ (short for $\mathbf{F}_{air\,on\,body}$). The body's only long-range interaction is its gravitational interaction with the earth, so the third force is the gravitational force $\mathbf{F}_g$ (short for $\mathbf{F}_{earth}$, itself short for $\mathbf{F}_{earth\,on\,body}$).

5.4.1 Without Air Resistance

Before we determine these forces and despite my harping on the importance of the air, first consider the simpler situation of bicycling without air resistance.

This step offers two benefits. First, it reveals the deep similarities underneath many superficially different examples of steady motion. This composite body, when freed of air resistance, experiences only two forces, the gravitational force and the contact force (Figure 5.4a): the same forces that you experience standing on level ground (Section 5.1), on a hill (Section 5.2), or in a steadily descending elevator (Section 5.3). In each case, the body's acceleration is zero, so the net force must also be zero. Because one force is mg downward, the second and balancing force (the contact force) must be mg upward. These diverse situations are therefore identical, despite the differing velocities and ground angles.

Second, this step continues the long process of unlearning the deepest misconception about force: the idea of a "force in the direction of motion" or "force of motion." The process started in Section 4.1 with the idea that force *changes* motion. Even so, in explaining a body's motion, one is tempted – the impersonal "one" conceals my own feeling of temptation – to invoke a force in the direction of motion (Figure 5.4b). For in our psyche, in which millions of years of evolution have incorporated millions of years of everyday experience, we deeply believe that force causes motion, that $\mathbf{F}_{net} \to m\mathbf{v}$.

However, as we deduced from reasoning about the composite body's only two interactions, the freebody diagram contains no horizontal forces at all, let alone a force in the direction of motion. Thus, force cannot cause motion directly. Instead, as we can remind ourselves, force *changes* motion.

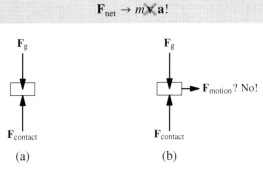

$$\mathbf{F}_{net} \to m\cancel{\times}\mathbf{a}!$$

(a)   (b)

Figure 5.4 The freebody diagram of the composite body (you and the bicycle) with no air resistance. (a) The composite body participates in two interactions and therefore experiences two forces. The body isn't accelerating, so the two forces must balance. (b) The body's motion creates a strong temptation to include a "force of motion" in that direction. However, this force belongs to no interaction. Thus, it doesn't exist. Rather, it's a relic of our ancient misconception that $\mathbf{F}_{net} \to m\mathbf{v}$.

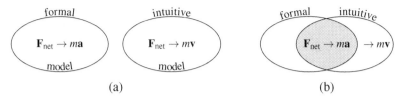

Figure 5.5 Aligning intuitive and formal models. (a) When we start studying Newton's laws, these models are completely separate. (b) Studying Newton's laws well
means merging the models. They cannot merge fully because $F_{net} \to mv$ is still
needed to function in the everyday world.

Studying Newton's laws effectively means revising our intuitive, $F_{net} \to mv$
model so that it partly merges with the formal, Newtonian, $F_{net} \to ma$ model.
For when the models are completely separate (Figure 5.5a), our intuition cannot
guide our formal reasoning. Our formal reasoning then degenerates into formalist
reasoning. We wander in the space of possible equations as we hope, usually
in vain, that this mostly random walk will reach a reasonable result. And even
when it does, we still won't understand why.

If, instead, we study Newton's laws reflectively, by diagnosing and trying to
undo our deep misconceptions, then our intuitive and the formal models merge.
They won't and shouldn't merge fully: The intuitive belief that $F_{net} \to mv$ is still
needed to function in the everyday world. Remember the hominids of Figure 4.1
who hoped to escape the hungry tiger with only a single push. Although the
tiger's descendants might still be with us, the hominids' descendants are not!
Ideally, our intuitive model becomes a mixture of $F_{net} \to mv$ and of $F_{net} \to ma$.
Its $F_{net} \to mv$ part guides our interactions with the everyday world as we open
sticky doors, push boxes across the floor, and run from tigers. And its shared
part, $F_{net} \to ma$, can guide our formal reasoning (Figure 5.5b).

5.4.2 With Air Resistance

After that philosophical and metacognitive analysis, let's now restore air resistance, the third force on the composite body, and determine the three forces'
rough magnitudes and directions. After we estimate each force, it'll go onto the
freebody diagram that we'll build stepwise (Figure 5.6).

1. *The gravitational force* F_g. If the composite body has a total mass of, say, 70
   kilograms or roughly 150 pounds (considered as a mass unit) – it's a light,
   high-tech bicycle – then the gravitational force, whose magnitude is given
   by (1.12), is 700 newtons downward. As a body force, its tail goes at the
   (composite) body's center of mass (Figure 5.6a).

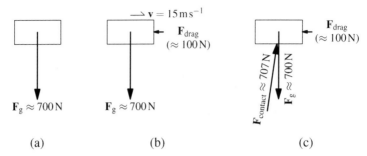

(a)   (b)   (c)

Figure 5.6 Building the freebody diagram for bicycling on level ground at constant speed. (a) The first force to include is gravity (the easiest of the three). (b) Then we include drag (the second easiest). As a reminder of the reason that there is drag, we indicate the body's velocity with an arrow (it has a single-sided harpoon to distinguish it from a force arrow). (c) Finally, we include the contact force, which is determined by the requirement that it balance the sum of gravity and drag.

2. *The drag force* $\mathbf{F}_{\text{drag}}$. Its magnitude is given by (1.21). Here, ρ is the density of air (about 1 kilogram per cubic meter), v is the body's speed (15 meters per second), and A_{cs} is the body's cross-sectional area.

   This cross-sectional area does include the cross-sectional area of the bicycle, but most of the cross-sectional area comes from you. To bicycle fast, you probably crouch in a racing position and span perhaps only 1 meter vertically. A typical person's width is about 0.5 meters. Then A_{cs} is about 0.5 square meters (Figure 5.7).

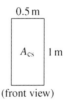

0.5 m

A_{cs} 1 m

(front view)

Figure 5.7 The composite body's cross-sectional area (front view). Imagine you and the bicycle coming out of the page, which tears. The area of torn paper is the composite body's cross-sectional area. This area is roughly 0.5 square meters.

With these values, F_{drag} is approximately 100 newtons:

$$F_{\text{drag}} \sim \underbrace{1 \text{ kg m}^{-3}}_{\rho} \times \underbrace{(15 \text{ m s}^{-1})^2}_{v^2} \times \underbrace{0.5 \text{ m}^2}_{A_{\text{cs}}} \approx 100 \text{ N}. \tag{5.6}$$

It points in the opposite direction to your motion: If you bicycle to the right, drag points to the left. As a contact force, it lies outside the composite body with its tip or tail at the surface.

Air is an extended object, so where do you place the tip or tail?

Drag's placement is indeed tricky. However, because drag is such a common force (alas!), its placement is worth clarifying. The rule for a contact force is to place its tip or tail at the point of contact: the tip when the contact interaction is repulsive, the tail when the interaction is attractive.

Drag arises from zillions of contact interactions over a body's entire surface. These interactions are all repulsive. Thus, to be complete, we should draw zillions of tiny arrows, all with their tips at the surface and pointing inward. (Except at minuscule speeds or for minuscule bodies, these little drag forces are pressure forces. These forces, as we'll discuss in Section 5.7, point perpendicularly inward.)

But such completeness is tedious. As the first simplification, forget about the forces on the side surfaces. Second, combine all the tiny forces on the front surface into one inward force at the center of the front surface. (Placing it at the center of the surface is a useful though not infallible rule of thumb.) Third, do the same for the back surface (Figure 5.8).

$\mathbf{F}_{\text{drag on back}}$ ⟶ ⟵ $\mathbf{F}_{\text{drag on front}}$

Figure 5.8 Lumping the infinity of tiny drag forces acting all over the body into two net drag forces, on the front and the back surfaces.

Finally, combine the two forces into one backward-pointing force on either the front or the back surface (Figure 5.9). I prefer to use the front surface because that placement follows the convention that a repulsive contact interaction puts the *tip* of its force arrow on the surface. Thus, in answer to the triangle question, point the drag force into the body, and place its tip at the center of the body's front surface.

$\mathbf{F}_{\text{drag}}$ ⟵ $\mathbf{F}_{\text{drag}}$ ⟵

(a)　　　　　　　　(b)

Figure 5.9 Two possibilities for placing a single, net drag force pointing to the left. (a) The preferred option: on the front surface. The *tip* of the arrow is then on the surface, correctly implying a repulsive contact interaction. (b) On the back surface.

With that choice, the freebody diagram now has two forces (Figure 5.6b). Because one of the forces, the drag force, depends on the body's velocity, the freebody diagram, for completeness, includes an arrow representing the body's velocity. This arrow is barbed only on a single side so that it is easily distinguished from a force arrow.

3. *The contact force from the ground,* $\mathbf{F}_{\text{contact}}$. This force, a passive force, comes last because it's the hardest to draw. In contrast to the gravitational force, which obviously points downward, and to the drag force, which obviously points backward, the contact force's direction is far from obvious. And how large is it? Both puzzles find their answer in Newton's second law. With zero acceleration, the net force must be zero:

$$\mathbf{F}_{\text{net}} = \mathbf{F}_{\text{g}} + \mathbf{F}_{\text{drag}} + \mathbf{F}_{\text{contact}} = 0. \tag{5.7}$$

Adding these three vectors tip-to-tail produces a triangle (Figure 5.10): $\mathbf{F}_{\text{g}}$ and $\mathbf{F}_{\text{drag}}$ are perpendicular and form two legs of this right triangle, and $\mathbf{F}_{\text{contact}}$, its hypotenuse, closes the triangle and makes the sum zero.

Thus, the contact force points upward and slightly in the direction of motion (to the right). To find its magnitude, we use the Pythagorean theorem:

$$F_{\text{contact}} = \sqrt{(700\,\text{N})^2 + (100\,\text{N})^2} \approx 707\,\text{N}. \tag{5.8}$$

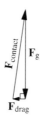

Figure 5.10 Determining the contact force. The net force must be zero, so the forces, added tip-to-tail, must form a closed path.

In answer to the triangle question (with drag): Now the freebody diagram shows the forces on the composite body (Figure 5.6c).

The contact force has three interesting features. First, although its vertical and horizontal components are 700 newtons and 100 newtons, respectively, its magnitude is almost exactly 700 newtons – as if the horizontal component were zero. This feature is a surprising consequence of the squaring and square-rooting operations in the Pythagorean theorem. Thus, when you must hurry, cross a field diagonally rather than going along two of the edges. Second, the contact force isn't perpendicular to the ground. Third, its nonperpendicular or parallel portion points in your direction of motion – rather than opposite to it.

▶ *The contact force's perpendicular portion is the familiar normal force* **N**, *but what's its parallel portion called?*

It is unlikely to be dynamic friction (Section 1.3.4), which, like drag, opposes your motion. Thus, it must be static friction (Section 1.3.5). This argument comes with two caveats. First, the bicycle also experiences rolling resistance

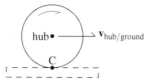

Figure 5.11 A bicycle wheel rolling to the right. The wheel touches the ground at point C. The hub has velocity $v_{hub/ground}$ (relative to the ground).

(caused, for example, by a partly deflated tire). Like dynamic friction, rolling resistance opposes your motion. However, it is usually small, so let's ignore it and its complications. Second, dynamic friction need not oppose your motion: It need only oppose the motion of the tire. (I soon return to this point, on p. 64.)

Meanwhile, keep the physics separate from the names. In spite of the two names, normal force and static friction, the ground applies only one force. This force's perpendicular portion balances gravity. Its parallel portion balances drag.

How can the bicycle wheel, which is constantly in motion, experience static friction, which implies no motion?

The point on the wheel where it touches the ground – point C in Figure 5.11 – does not move relative to the ground. To understand this strange fact, we can decompose the velocity of point C into two portions: (1) $v_{hub/ground}$, which points forward (the wheel's hub moves forward relative to the ground), and (2) $v_{C/hub}$, which points backward (point C moves backward *relative to the hub*). In the subscripts, the slash / means "relative to" or "in the reference frame of" (with the understanding that the hub reference frame merely translates and does not rotate). Then, rearranging and restating (3.5) symbolically,

$$v_{C/ground} = v_{C/hub} + v_{hub/ground}. \qquad (5.9)$$

For most bicycling, $v_{C/hub} = -v_{hub/ground}$. In this common situation where the two velocities balance, $v_{C/ground}$ is zero, meaning that the bottom of the wheel does not move relative to the ground and can experience static friction.

Why do the two velocities usually balance?

Here a thought experiment helps. Imagine that $v_{hub/ground}$ is slightly greater than $v_{C/hub}$: The wheel isn't spinning quite fast enough. Then point C, the bottom of the wheel, moves forward slowly relative to the ground. Thus, dynamic friction points backward, applying to the wheel a clockwise twist (technically, a torque) – which, as I'll explain in Section 8.2, tries to *increase* the wheel's spin and therefore also $v_{C/hub}$. When the spin increases enough to make $v_{C/hub}$ equal $v_{hub/ground}$, point C no longer moves relative to the ground. Dynamic friction vanishes, and static friction returns. In short, negative feedback from the ground usually keeps the two velocities balanced.

Extending this thought experiment shows why the velocities don't always balance. Imagine bicycling with $v_{hub/ground}$ equal to $v_{C/hub}$ but then braking so hard that the brake shoes clamp hard on the rim and stop the wheel from spinning. Braking makes $\mathbf{v}_{C/hub}$ zero. Thus, from (5.9), $\mathbf{v}_{C/ground}$ equals $\mathbf{v}_{hub/ground}$ and is nonzero: You're skidding, and the tires experience dynamic friction.

For the mirror-image thought experiment, apply a huge force to the pedals, enough to spin up the wheels quickly and make $v_{C/hub}$ greater than $v_{hub/ground}$. As the gearheads say, you're burning rubber. In this case, dynamic friction points in the same direction as your motion (but still opposite to the motion of the tire) – the second caveat that I mentioned on p. 62.

In most bicycling, however, when you neither skid nor burn rubber, point C does not move relative to the ground, and the bicycle experiences static friction.

5.5 Sledding at Constant Velocity

Now we can extend this analysis to a situation with nonhorizontal motion and an unknown speed. It happens to me in the New England winter: I sled down a neighborhood hill at constant velocity (Figure 5.12). To enliven the sledding and simplify the trigonometry, this hill has a 30-degree inclination or tilt. (A lower, more realistic slope is the subject of Problem 5.2.) Our goal is to estimate my sledding speed: Roughly how fast do I zoom down the hill?

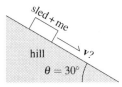

Figure 5.12 Sledding downhill at constant velocity. How fast does the composite body (of the sled and me) zoom down the hill?

5.5.1 Without Dynamic Friction

To declutter the analysis, first pretend that the sled is perfectly waxed, a pretense that eliminates dynamic friction between the sled and the snow. (Friction returns in Section 5.5.2.) The analysis is, in the terminology of Section 5.1, a mix of type I (inferring) and type C (calculating). We know something about the forces on me and the sled (for example, we know the gravitational force) and something about my velocity (we know its direction and that it's not changing) but need to complete these areas of knowledge consistently.

To do so, we find all the forces on the composite body of me and the sled (henceforth called me), require that the forces combine to produce zero net force – the only way to produce zero acceleration – and see what that requirement implies for my speed.

These forces belong on a freebody diagram, which we make by first finding all my interactions. My only long-range interaction is the gravitational interaction with the earth. My short-range (contact) interactions are with the two bodies that I touch: the hill and the air. The sled belongs to the composite body known as me, so it doesn't provide another contact force. (Actually, the sled–me interaction provides two contact forces. However, because of Newton's third law, they cancel out in the net-force sum. This important point is discussed in Section 7.2.)

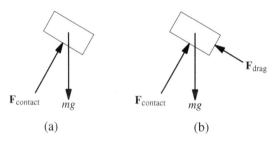

Figure 5.13 Drawing the forces on me stepwise. (a) Drawing gravity and the contact force. Without dynamic friction, the contact force is perpendicular to the hill (its magnitude is still unknown). (b) Drawing the drag force. Its direction is also known (opposite to my motion), but its magnitude is still unknown.

Each of the three interactions contributes one force acting on me.

1. *The gravitational force,* $\mathbf{F}_g$. It points down and has magnitude mg, where m is my mass (including the sled).

2. *The ground's contact force,* $\mathbf{F}_{contact}$. I slide relative to the ground and don't experience static friction; and, because of the perfect wax, I experience no dynamic friction. Friction is the parallel portion of the contact force. Thus, this contact force has no parallel portion and points perpendicularly to the hill (Figure 5.13a). This requirement determines the contact force's direction; its magnitude will be determined by the condition of zero net force.

3. *The air's contact force,* $\mathbf{F}_{drag}$. This force opposes my motion, so it points parallel to and up the hill (Figure 5.13b). This constraint determines $\mathbf{F}_{drag}$'s direction. According to (1.21), its magnitude F_{drag} depends on my speed v, which is still unknown. But F_{drag} will be determined by the condition of zero net force – and F_{drag} will then determine my speed (our goal).

To find the two unknown magnitudes, we use the information that my acceleration is zero. Acceleration is produced by net force – the first crucial idea of the second law (Section 4.1). Therefore, we know that the net force is also zero.

▶ *Why do I fussily insist that, "Therefore, we know that the net force is zero," rather than simply saying, "Therefore, the net force is zero"?*

I want to separate the direction of this argument from the direction of physical causality. The argument operates in our mental world. There, our knowing the net force does cause us to know the acceleration. However, in the physical world, the causation operates in the opposite direction: Force causes acceleration.

Mistaking our logic for the world's logic is to commit the mind-projection fallacy [9]: attributing to the world what is instead a property of our thinking. Thus, avoid saying, "Because the sled isn't accelerating, the net force on it is zero." The "be*cause*" implies a causal physical relationship pointing in the wrong direction: It implies that the sled's lack of acceleration is the cause and that the zero net force is the result (or effect). Instead, try to say and think, "Because the block isn't accelerating, *I know* that the net force on it is zero."

To describe the causation fully, I even try to say, "Because *I know* that the block isn't accelerating, *I know* that the net force on it is zero." Now the cause-and-effect relationship is properly described as lying in the mental world. The cause is our realizing that the block isn't accelerating. The effect of this cause is our realizing that the net force is zero.

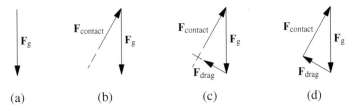

(a)   (b)   (c)   (d)

Figure 5.14 Finding the magnitudes of the drag and contact forces graphically. (a) Draw the gravitational force. (b) Add the contact force tip-to-tail in the known direction but with a dashed tail indicating that the magnitude isn't yet known. (c) Add the drag force with a dashed tip. (d) Chop off these forces' extensions beyond their intersection. Now the magnitudes are known (after some trigonometry).

With that epistemological sermon over, let's return to finding the forces and, in particular, to requiring that the net force be zero. The analysis, a construction, has four steps (Figure 5.14). It turns on the idea that, for the net force to be zero, the three forces, when added by drawing them tip-to-tail, form a triangle.

1. Start the triangle by drawing the gravitational force – the one force whose direction and magnitude we already know.

2. Next, draw the contact force in the correct direction, perpendicular to the hill. Its tip lies at the tail of $\mathbf{F}_g$, making the sum so far $\mathbf{F}_{contact} + \mathbf{F}_g$. Its dashed tail indicates that we don't (yet) know its magnitude.

3. Then, draw the drag force in the correct direction (up the hill). Its tail lies at the tip of $\mathbf{F}_g$, making the sum $\mathbf{F}_{contact} + \mathbf{F}_g + \mathbf{F}_{drag}$, which is the net force. Its dashed tip indicates that we don't (yet) know its magnitude.

4. Finally, require that the net force be zero. Pictorially, the tip of $\mathbf{F}_{drag}$ (the final arrow in the sum) must lie at the tail of $\mathbf{F}_{contact}$ (the first arrow in the sum). Then the intersection of the extended drag- and contact-force arrows determines each force's magnitude.

Because the drag force points along the hill and the contact force points perpendicularly to the hill, these two forces are perpendicular. Thus, the force-sum triangle is a right triangle. Its hypotenuse, the gravitational force, has length mg. The two leg lengths can be found from trigonometry using the two nonright angles. For this wild sled ride, they are 30 degrees and 60 degrees.

But which nonright angle has which value?

The simplest way to decide is to use the method of easy cases. Mentally flatten the hill, shrinking its inclination θ from 30 degrees toward zero (Figure 5.15a). As θ shrinks, the contact force aligns ever more closely with the gravitational force (though pointing in the opposite direction). Thus, the triangle's top angle (the smallest angle) approaches zero – which, in this easy case, is the hill's inclination θ. Thus, this top angle, in the actual case of a 30-degree inclination, is also 30 degrees (Figure 5.15b). The method of easy cases is fantastic. (I have written much, perhaps too much, about it in *Street-Fighting Mathematics* [13, Chapter 2] and *The Art of Insight in Science and Engineering* [14, Chapter 8].)

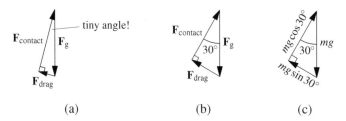

(a) (b) (c)

Figure 5.15 Finding the angles and sides using the easy case $\theta \to 0$. (a) Shrinking the hill's inclination θ toward zero. Then the top angle also shrinks. (b) Thus, the top angle must be θ (rather than $90° - \theta$), which is $30°$. (c) The legs of this right triangle can now be found using trigonometry.

The leg lengths are the magnitudes of the forces. Thus, the contact force has magnitude $mg \cos \theta$ or, here, $mg \cos 30°$ (Figure 5.15c). With a mass m of 70 kilograms (me plus sled), mg is 700 newtons. With $\cos 30° = \sqrt{3}/2$,

$$F_{\text{contact}} = mg \cos 30° \approx 600\,\text{N}. \tag{5.10}$$

Because the force is perpendicular to the surface of contact, it's also called the normal force and labeled **N**. (You don't actually need to determine N in this first analysis with zero dynamic friction. However, you need to do so once dynamic friction returns because its magnitude is proportional to N.)

Similarly, and using $\sin 30° = 0.5$,

$$F_{\text{drag}} = mg \sin 30° = 350\,\text{N}. \tag{5.11}$$

Knowing F_{drag}, we can find my speed.

▶ *At what speed do I have to sled to produce that much drag?*

This speed, called the terminal speed, is connected to F_{drag} through (1.21). With $\rho \approx 1$ kilogram per cubic meter, $A_{\text{cs}} \approx 0.25$ square meters (I lie flat on my stomach, praying for a safe ending), and $F_{\text{drag}} = 350$ newtons, my speed is roughly 37 meters per second (roughly 130 kilometers, or 80 miles, per hour):

$$v \sim \sqrt{\frac{F_{\text{drag}}}{\rho A_{\text{cs}}}} \approx \sqrt{\frac{350\,\text{N}}{1\,\text{kg}\,\text{m}^{-3} \times 0.25\,\text{m}^2}} = \sqrt{1400}\,\text{m}\,\text{s}^{-1} \approx 37\,\text{m}\,\text{s}^{-1}. \tag{5.12}$$

In any unit system, it's dangerous sledding.

5.5.2 Adding Dynamic Friction

Now let's restore a bit of realism by including a small amount of dynamic friction (Section 1.3.4), taking the dynamic-friction coefficient μ to be 0.3. This value is high for the slippery ice and snow and any half-decent sled, but I use a deliberately roughened sled surface – an attempt to slow my sledding to a less frightening (but constant) speed.

With dynamic friction present, the contact force no longer points perpendicularly to the hill. Rather, a small portion of it, namely dynamic friction, lies along the hill. Its magnitude, from (1.26), is μN, where N is the magnitude of the normal force – which itself is the contact force's perpendicular portion. This portion, **N**, is the same as it was without dynamic friction, for it must still balance the perpendicular portion of the (unchanged) gravitational force. No other perpendicular forces are available to do that job. Thus, N still equals $mg \cos 30°$ or roughly 600 newtons.

If you frequently break the contact force into its parallel and perpendicular portions, why do you still emphasize the contact force?

The emphasis recalls a key principle of Newton's laws: Each force belongs to one interaction. Equivalently, each interaction in which a body participates contributes one force on the body. Thus, my (really, the sled's) interaction with the ground contributes exactly one force, the contact force.

Knowing N, we can calculate the dynamic-friction magnitude:

$$F_\mu \approx \underbrace{0.3}_{\mu} \times \underbrace{600\,\text{N}}_{\approx N} = 180\,\text{N}. \tag{5.13}$$

Thus, now there are four force vectors: gravity, drag, dynamic friction, and the normal force – with the last two being portions of a single contact force. Pictorially, the three sides of the right triangle keep their frictionless lengths and orientations, but the short leg splits into two portions: drag and dynamic friction (Figure 5.16a). These two portions together balance the parallel portion of gravity (leaving the normal force and the perpendicular portion of gravity to balance each other, as they did without dynamic friction).

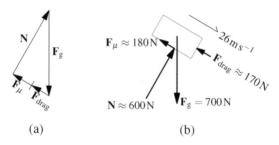

(a)   (b)

Figure 5.16 The revised forces when dynamic friction is restored. (a) The force sum. The drag force and the new dynamic-friction force ($\mathbf{F}_\mu$) now together balance the parallel portion of the gravitational force. (b) The new freebody diagram with the calculated forces.

Because both forces point in the same direction (uphill), the drag is smaller than it was without dynamic friction. Now F_{drag} is approximately $(350 - 180)$ newtons or 170 newtons (Figure 5.16b). My speed is then roughly 26 meters per second (roughly 100 kilometers, or 60 miles, per hour):

$$v \sim \sqrt{\frac{170\,\text{N}}{1\,\text{kg m}^{-3} \times 0.25\,\text{m}^2}} = \sqrt{680}\,\text{m s}^{-1} \approx 26\,\text{m s}^{-1}. \tag{5.14}$$

It's slower than without dynamic friction but still too fast for my blood.

5.6 Tension and Tension Forces

We next investigate a common "force" in the everyday world, particularly in situations of zero acceleration: tension.

What is tension? Is it a fifth force beyond the four kinds in Section 1.2.1?

The classification might well have omitted a fifth kind of force – the universe is mysterious and our understanding limited – but tension is not such a force. For tension isn't even a force, despite its frequent portrayal as one. For example, one often hears that a person or a weight "was pulled upward by the tension in the rope." This misinterpretation reflects and causes many confusions.

Instead, tension describes a state within a body (typically, a thin solid like a string or a rod) that produces a contact interaction and, therefore, contact forces at each end of the body. These forces are electromagnetic, one of the four kinds of force. Thus, don't confuse tension, an internal state, with these tension forces – which I also call *forces due to tension* and label $\mathbf{F}_T$ (not T). Tension forces are actually spring forces (Section 1.3.2), but the extension of the spring is usually too small to see or is idealized away. For example, in physics, a string is defined as a massless, flexible, and inextensible connection. Thus, no matter its tension, its extension is always zero.

5.6.1 Tug-of-War

To show how tension forces work, we'll use them to make freebody diagrams of a system with two people, me and you, playing tug-of-war (Figure 5.17). We each hold one end of a string. (A string is physics code for a rope with no mass.) In keeping with the ground rules in this chapter, neither of us is accelerating ($\mathbf{a} = 0$). In keeping with how tug-of-war is mostly played, we are not moving at all ($\mathbf{v} = 0$ in addition to $\mathbf{a} = 0$): We are planted solidly on the ground and fighting to a draw. For simplicity, we have the same mass m.

The system contains four bodies: you, me, the string, and the earth. Let's consider a plausible candidate for my freebody diagram (Figure 5.18a).

What's wrong with this freebody diagram?

To decide, use Newton's second law: Net force causes acceleration. Here, even if $\mathbf{N}$ balances the gravitational force, a net force remains in the horizontal direction (due to $\mathbf{F}_{string}$). Thus, I would accelerate – contrary to the assumption that we tug to a draw. So, this diagram cannot be correct.

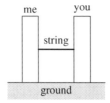

Figure 5.17 Playing tug-of-war. You and I each pull on the string (a massless rope) trying to topple the other. We are tugging to a draw: Neither of us is accelerating ($\mathbf{a} = 0$) or moving ($\mathbf{v} = 0$).

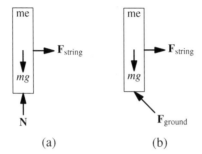

Figure 5.18 Freebody diagrams of me while playing tug-of-war. (a) An incorrect diagram. (b) The fixed diagram.

We can fix it using first principles: the recipe for making a freebody diagram (Section 2.1). Because the long-range interactions here are simpler than the short-range interactions, we can modify the recipe and handle the long-range interactions first to get them out of the way and free up working memory for the short-range interactions.

My only long-range interaction is the gravitational interaction with the earth. This interaction contributes the gravitational force, mg downward.

Meanwhile, I have two short-range interactions, one for each body that I touch: the ground (part of the earth) and the string. Each short-range interaction contributes one contact force on me. The contact force of the string correctly points horizontally, along the string. But the contact force of the ground need not point upward. It could have a portion parallel to the ground: static friction ("static" because I am planted solidly on the ground).

$\mathbf{F}_{ground}$, through its parallel portion (static friction), balances $\mathbf{F}_{string}$. Through its perpendicular portion (the normal force), it balances the gravitational force (Figure 5.18b). The consequence is zero net force and zero acceleration.

Thus – in answer to the triangle question – the wrong freebody diagram of me (Figure 5.18a) omitted the static-friction portion of the contact force, making the net force and my acceleration nonzero (contrary to the assumption that we are tugging to a draw).

Although Figure 5.18b correctly shows the forces, my stance is wrong. If I try to stand upright, as we all know from tug-of-war experience, my attempt quickly ends when the string pulls me over. This aspect of my motion, a rotation, cannot be analyzed with Newton's laws alone. It requires the ideas of angular momentum and torque, which you meet, albeit briefly, in Section 8.2 as part of "What comes next (after Newton's laws)." As a shortcut to avoid calculating, we can use tug-of-war experience: We all know to lean backward (Figure 5.19a).

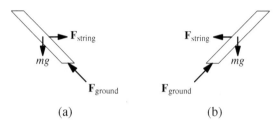

(a)   (b)

Figure 5.19 Leaning backward. (a) My revised freebody diagram. The forces are the same as in Figure 5.18b; however, to avoid face-planting, I lean backward. (b) You lean in the opposite direction.

▶ *What's your freebody diagram?*

Your diagram (Figure 5.19b) mirrors my diagram.

▶ *What's the freebody diagram of the string?*

Its diagram is tricky. Thus, we return to first principles: We list the string's interactions and, for each interaction, draw one force on the string.

▶ *In what interactions does the string participate?*

The string's only possible long-range interaction is the gravitational interaction with the earth. However, because it's a string, the physics code for a rope with zero mass, it does not participate in a gravitational interaction. Equivalently, the gravitational force is zero. Either way, there's no long-range force to draw.

To find the string's short-range (contact) interactions, we just ask ourselves which objects it touches. The answer: me and you. Each touch is an interaction. Thus, the string participates in two interactions, both short range.

Each interaction contributes one force on the string. Because the string isn't accelerating, you know that these two tension forces, each acting at one end of the string, must balance (Figure 5.20a).

▶ *Do the two forces balance because of the second law or the third law?*

Because this distinction is so easy to get wrong, return again to the first principle: interactions. The two forces in question belong to separate interactions. Thus, they cannot be connected by the third law (which connects the two forces of one interaction). Rather, the forces are connected by Newton's second law: by its requirement that zero acceleration requires zero net force. (Even if the string were accelerating, the two forces on it would still have to balance because it's massless. Otherwise, because of the $1/m$ factor in the second law (4.5), the string would get an impossible infinite acceleration.)

Although two forces that balance are also equal (in magnitude) and opposite (in direction), I use "balance" when the reason is the second law and "are equal and opposite" when the reason is the third law. This convention helps us to distinguish when the forces' relation is contingent and dependent on the body's acceleration (relevant for the second law) from when the relation always holds (a result of the third law).

Either tension force's *magnitude* F_T equals the tension T in the string:

$$F_T = T. \tag{5.15}$$

But neither F_T nor T is a vector! F_T is a magnitude. And, as a reminder, tension isn't a force itself. Rather, it's a condition in a material that produces contact forces at its ends. Thus, don't write $\mathbf{T}$ (a vector). It's too easily misinterpreted as the tension itself, reinforcing the misconception that tension is a force and therefore a vector.

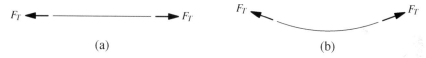

(a)   (b)

Figure 5.20 Freebody diagrams for a straight and a sagging tug-of-war string. (a) The correct freebody diagram for a horizontal string. (b) A plausible but incorrect freebody diagram for a sagging string. The net force on this string isn't zero, so it would have an impossible infinite acceleration.

Why do I draw the string perfectly horizontal? Couldn't it sag in the middle?

In my experience of tug-of-war, the rope does sag in the middle. However, a rope has mass. In contrast, the idealized rope known as a string has no mass.

If a tug-of-war string sagged, its freebody diagram would have two tension forces that point slightly upward (Figure 5.20b), because tension forces point along the string that produces them. If they did not, the string, being flexible, would bend until the string and the tension force do align.

These tension forces, no longer opposite in direction, cannot balance. Thus, the net force on the string would be nonzero. Because the string is massless, this net force would give it an infinite acceleration (acceleration is inversely proportional to mass). Nonsense! Therefore, the string must be horizontal. (In Problem 5.6, you generalize these freebody diagrams to the case of a rope.)

▶ *What's the freebody diagram of the earth?*

By now, you know the drill: Determine the earth's interactions. It participates in two long-range interactions: the gravitational interaction with me and the gravitational interaction with you. And it participates in two short-range interaction: the contact interaction between me and the ground (considered as a part of the earth) and the contact interaction between you and the ground. The gravitational interactions contribute two upward forces, each with magnitude mg. The contact interactions contribute two forces, each equal and opposite to the respective contact force on you or me (Figure 5.21).

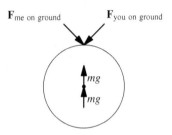

Figure 5.21 Freebody diagram of the earth. It experiences two gravitational forces, one from each of us, and two contact forces, again one from each of us.

▶ *Among these freebody diagrams of me, you, the string, and the earth, what are the third-law force pairs?*

Figure 5.22 shows the freebody diagrams with the third-law pairs connected by dashed lines. The two sides of an interaction never act on the same body, so each third-law pair reaches between two diagrams.

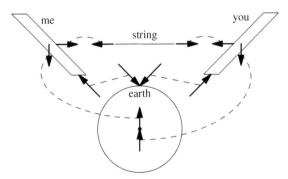

Figure 5.22 All four freebody diagrams of tug-of-war, with third-law pairs connected by dashed lines.

5.6.2 Mass Suspended with Three Strings

The next example combines tension with trigonometry. A block is suspended using three strings, imaginatively labeled 1, 2, and 3 (Figure 5.23). Our goal is to find the three string tensions T_1, T_2, and T_3. (In real life, we would make such a calculation to ensure that no tension exceeds what that string can withstand.)

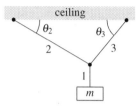

Figure 5.23 A stationary block suspended using three strings.

The string tension T_1 can be found by making the freebody diagram of the block (Figure 5.24). The block participates in two interactions: the long-range, gravitational interaction with the earth and the short-range, contact interaction with string 1. Thus, two forces act on the block. The gravitational interaction contributes a downward force with magnitude mg. The contact interaction contributes an upward tension force with magnitude F_{T_1}.

Figure 5.24 The freebody of the block. It participates in only one contact interaction, with string 1, so it experiences two forces (including the gravitational force).

Because we know that the block isn't accelerating (it's not moving ever), we know that the net force on it is zero. Thus, the contact force balances the gravitational force. So, $F_{T_1} = mg$. (As vectors, $\mathbf{F}_{T_1} = -m\mathbf{g}$.) The magnitude of a tension force is the tension itself ($F_{T_1} = T_1$), so $T_1 = mg$.

➤ *Why am I so fussy about distinguishing T_1 (the tension) from F_{T_1} (the magnitude of the tension force)?*

This important distinction reminds us that tension and force are different beasts. The distinction – introduced on p. 70 but so often forgotten that it merits another mention – escaped me for the first 10 years that I taught physics. To those students: I hereby ask your forgiveness for passing on my confusion.

To conflate tension and force is to commit a category mistake – perhaps the ultimate philosophical crime. A force, such as the tension force $\mathbf{F}_{T_1}$ on the block, is a push or a pull and is one side of an interaction. And F_{T_1} is that force's magnitude. In contrast, tension itself describes the state within a material (usually a solid) that produces such forces at each end. Thus, to indicate a contact force produced by a tension T, I write $\mathbf{F}_T$. Even though F_T, the magnitude of the tension force, equals the tension T in that they are the same number of newtons (or of any force unit), they belong to different physical categories.

Knowing that $T_1 = mg$, we can find the other two tensions. The subtle method is by making a freebody diagram of a point: the intersection of the three strings. In real life, which knows no points, the point is really a tiny knot. So, if you don't like thinking of the freebody diagram of a point, you can imagine instead a freebody diagram of a tiny knot.

This point or knot participates in three contact interactions, one with each string. However, it participates in no gravitational interaction: Whether imagined as a point or as a knot made from tiny pieces of massless strings, it's massless.

Thus, the knot experiences three forces: one tension force from each string. For two reasons, these forces must add to zero. First, the knot isn't accelerating. Therefore, the net force on it must be zero. Second, even if the knot were accelerating, the net force on it would still have to be zero because it's massless.

These tension forces added tip-to-tail (Figure 5.25) make a closed figure: the geometric statement of zero net force. Because there are three tension forces, the closed figure is a triangle. Its string-1 side has a known length (mg) and direction (downward). Its other two sides have known directions (specified through θ_2 and θ_3) but unknown lengths because T_2 and T_3 are still unknown.

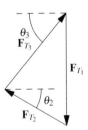

Figure 5.25 Adding the three tension forces acting on the knot. Because the knot isn't accelerating (or because it's massless), we know that the net force on it is zero. Thus, the tip-to-tail vector sum forms a triangle.

Solving for T_2 and T_3 is easy when the three sides make a right triangle. An example was the analysis of constant-speed, frictionless sledding (Section 5.5.1), where the gravitational force played the role of $\mathbf{F}_{T_1}$ here, drag played the role of $\mathbf{F}_{T_2}$, and the normal force played the role of $\mathbf{F}_{T_3}$.

▶ *But do the three tension forces make a right triangle?*

They do in special cases. When $\theta_2 + \theta_3 = 90°$, the T_2 and T_3 sides form a right angle. However, for general values of θ_2 and θ_3, the triangle isn't right.

Thus, finding the unknown lengths requires less familiar trigonometry: the law of sines (Figure 5.26a). In a triangle with sides a, b, and c and with opposite angles A, B, and C, the sides and angles are related by

$$\frac{a}{\sin A} = \frac{b}{\sin B} = \frac{c}{\sin C}. \qquad (5.16)$$

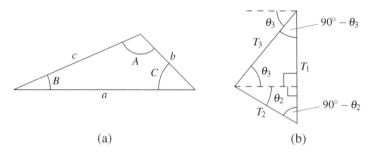

(a)   (b)

Figure 5.26 Applying the law of sines. (a) The law of sines (5.16) connects the sides (a, b, and c) to their respective opposite angles (A, B, and C). (b) Labeling the sides of the vector-sum triangle (Figure 5.25) with their respective magnitudes, the tensions, and labeling several angles to foster applying the law of sines.

The side lengths in the force triangle are T_1, T_2, and T_3 (Figure 5.26b). The opposite angles are, respectively, $\theta_2 + \theta_3$, $90° - \theta_3$, and $90° - \theta_2$. Therefore, from the law of sines (5.16),

$$\frac{T_1}{\sin(\theta_2 + \theta_3)} = \frac{T_2}{\sin(90° - \theta_3)} = \frac{T_3}{\sin(90° - \theta_2)}. \qquad (5.17)$$

Because $\sin(90° - \theta) = \cos\theta$, the last two denominators simplify to give

$$\frac{T_1}{\sin(\theta_2 + \theta_3)} = \frac{T_2}{\cos\theta_3} = \frac{T_3}{\cos\theta_2}. \qquad (5.18)$$

The T_1 ratio contains only known quantities: T_1 is mg, and the two angles are given. In terms of these known quantities, the three tensions are

$$T_1 = mg,$$
$$T_2 = mg\,\frac{\cos\theta_3}{\sin(\theta_2 + \theta_3)}, \text{ and} \qquad (5.19)$$
$$T_3 = mg\,\frac{\cos\theta_2}{\sin(\theta_2 + \theta_3)}.$$

Although T_2 and T_3 look intimidating, you can check that they make sense in a limiting, simpler case – try Problem 5.3.

5.6.3 The Atwood Machine with Equal Masses

Our final tension example helps you avoid a common and natural misconception: that statics, the topic of this chapter, requires that bodies be *static* in the word's everyday meaning of unchanging or stationary. In statics, the net force on any body is zero and so is any body's acceleration. However, the body can have any velocity, as long as the velocity does not change – the meaning of zero acceleration. You have seen a few examples of motion with constant but nonzero velocity: standing in a descending elevator (Section 5.3), bicycling on level ground (Section 5.4), and sledding at constant speed (Section 5.5). However, the principle – the distinction between velocity and acceleration – is so important and so often ignored that it deserves another example.

The example is a special case of a famous device, the Atwood machine (invented in 1784 by George Atwood). In idealized form, the machine consists of two masses hanging from a connecting string looped around a frictionless pulley (Figure 5.27a). The machine reduced the effect of the earth's gravity in a predictable way, so that gravity's full strength – the value of g – could be determined accurately using technology of the 18th century.

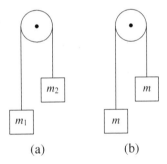

(a)   (b)

Figure 5.27 The Atwood machine. (a) The general case, usually with $m_1 \neq m_2$. (b) The special case of equal masses.

The analysis of that system, where the two masses are not quite equal, lies outside the scope of this chapter with its limitation to zero acceleration. Here, the two masses are equal (Figure 5.27b); let's also ignore air resistance. But even with that restriction, is zero acceleration even possible without zero velocity?

When the masses are equal, can the left mass rise (or fall) with a constant nonzero speed while the right mass falls (or rises) with the same speed? Or, once the masses are moving, must they accelerate?

To decide, assume that the masses move with constant velocity, make their freebody diagrams, and see whether the diagrams make consistent predictions.

Each mass participates in two interactions: one short-range interaction, the contact interaction with the string, and one long-range interaction, the gravitational interaction with the earth. The contact interaction contributes a tension force F_T upward (where F_T, the magnitude of the tension force, equals the string tension T). The gravitational interaction contributes a gravitational force mg downward (Figure 5.28).

Figure 5.28 The freebody diagram of either mass. Each mass participates in two interactions and therefore experiences two forces.

Because each mass moves at constant velocity (the starting assumption), the net force on each mass must be zero – a deduction made by inverting the causation from force to acceleration. Thus – and in answer to the triangle question – as long as the tension force balances gravity, meaning that $F_T = mg$ or that the string tension T remains mg, the masses move at a constant nonzero velocity.

The requirement that T remain mg means that the string can neither stretch (which, because strings are inextensible, would require an infinite tension) nor shrink (which would release the tension, making T zero). Thus, the constant descent speed of one mass equals the constant ascent speed of the other.

They move at constant velocity even though the net force on each mass is zero. Like me, you might find this conclusion surprising, which means that our intuitions don't fully agree with Newton's second law. We can bring our intuitions closer to the second law by making our intuitions explicit and comparing them to the second law.

For example, I can intuitively and easily imagine the masses at rest. I can also easily imagine the lower mass (in Figure 5.27b, the left mass) descending while the higher mass ascends. But I cannot easily imagine the lower mass ascending while the higher mass descends. That motion combination just feels asymmetric, and any asymmetry feels unwarranted: With equal masses, there just shouldn't be any asymmetry.

That intuition, alas, uses the deeply rooted misconception that force causes velocity (rather than causing acceleration). For only with velocity as the yardstick is there any asymmetry. In contrast, with acceleration as the yardstick, the system is symmetric. The acceleration of either mass is zero, whether the mass is descending, at rest, or ascending. And zero is the most symmetric value. Thus, in the Newtonian world picture, where force causes acceleration, the system behaves properly, even when the lower mass ascends. I remind myself of this example whenever I am tempted to think that force causes velocity!

◀ *Can unequal masses (Figure 5.27a) move at constant velocity?*

Constant-velocity or zero-acceleration motion requires that, on each mass, the tension force balance the gravitational force. Thus, the tension varies from $m_1 g$ at the left end of the string to $m_2 g$ at the right end. However, in a frictionless string or in a string on a frictionless surface, the tension must be the same throughout the string (what you show in Problem 5.9). Thus, when the masses are unequal, constant-velocity motion is impossible: The masses must accelerate.

Indeed, as I mentioned in introducing the Atwood machine (p. 78), measuring their small but nonzero acceleration (when m_1 is close but not equal to m_2) is a way to measure g accurately and the reason that Atwood developed his machine. Calculating that acceleration is the subject of Problem 7.4 and requires using the second law with nonzero acceleration (the topic of Chapter 7).

5.7 Pressure versus Depth in a Lake

For the final example of zero-acceleration motion, we find the pressure versus depth in a lake (or in a swimming pool). I've chosen a lake as our example rather than an ocean because a lake is shallow enough that the water density hardly changes. However, before we find the pressure, you might well wonder:

◀ *What is pressure?*

This question raises yet another point that confused me for decades. For pressure is often described as force divided by (or "per") area. But that description is at best half correct. It correctly gives the dimensions of pressure, from which come its units. In the metric (SI) system, the units are newtons per square meter ($N\,m^{-2}$), also called pascals (Pa). However, the "force divided by area" description incorrectly implies that pressure, like force, is a vector.

Pressure isn't a vector. Rather, it's "a condition in a fluid that causes forces to act normally [perpendicularly] to every surface" [25, p. 31]. These pressure forces (Figure 5.29) point into the surface and have magnitude proportional to the pressure p and to the surface's area A. Pressure forces, like tension forces, are short-range electromagnetic contact forces. However, pressure itself, like tension itself, is neither a force nor a vector.

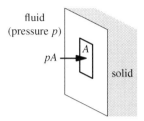

Figure 5.29 Pressure forces. The fluid, to the left of the surface, has pressure p. A piece of the surface with area A experiences a pressure force that points into and perpendicular to the surface and has magnitude pA.

If pressure isn't a vector, what is it?

Pressure is a pseudoscalar. All scalars, such as temperature, volume, and energy, have magnitude but not direction. Scalars come in two classes depending on their dimensions. *True scalars* contain length to an even power. Examples include temperature (which contains length to the zeroth power) and energy (which contains length to the second power). *Pseudoscalars* contain length to an odd power. Examples include volume (which contains length to the third power), tension (which contains length to the first power), and pressure (which contains length to the minus first power).

I learned these distinctions from J. W. Warren's discussion [25, App. 1], which also distinguishes true vectors, which contain length to an odd power, from pseudovectors, which contain length to an even power. Force is a true vector. Surprisingly, area is not a scalar but rather a pseudovector. It has magnitude equal to the area of a surface and points perpendicularly to the surface.

This view of area along with the informal description that "pressure is force divided by area" make it plausible that pressure is a pseudoscalar. Pressure is then a true vector (force) over a pseudovector (area). With one "pseudo" in the fraction, the quotient is plausibly "pseudo." And a vector over a vector is plausibly a scalar. Thus, the result is plausibly a pseudoscalar.

Now we're ready to find the pressure in the lake. To do so, we make a freebody diagram of a tube of water. The tube starts at the surface and descends into the lake to a depth h (but not all the way to the bottom of the lake).

To make the freebody diagram, the first step is to identify the tube's contact interactions. Here, the tube touches the air above it and the rest of the lake around and below it. However, the analysis simplifies if we split the rest of the lake into two bodies of water (Figure 5.30): (1) water down to the bottom of the tube and (2) water deeper than the tube. Then the tube touches three bodies, participates in three contact interactions, and experiences three contact forces.

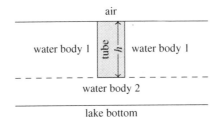

Figure 5.30 A tube of water in the lake. Above the tube sits air. Below it lies water down to the bottom of the lake (water body 2). And surrounding it is water at the same depth (water body 1, shown in two parts in this cross-sectional view).

Among the forces, the contact force due to water body 1 is the easiest to determine: It must be zero. If it were not zero, in which direction should it push the tube? Any particular direction breaks symmetry for no cause – which makes "no direction" the only answer. Thus, $\mathbf{F}_{\text{water body 1}} = 0$.

The remaining contact forces act on the tube's top and bottom surfaces. On the top surface sits the air (the atmosphere). At sea level, air pressure is $p_0 \approx 10^5$ pascals. Thus, on the tube's top surface, the pressure force points downward and has magnitude $p_0 A$, where A is the area of the tube's top surface. Under the bottom surface sits the lake water (water body 2). Its pressure right at the tube's bottom surface is $p(h)$ – an as-yet-unknown function of depth. The pressure force on the tube's bottom surface has magnitude $p(h)A$ and points upward.

➤ *Are these contact forces the only forces acting on the tube?*

No! In my rush to handle the contact interactions and apply the sly symmetry argument, I forgot one interaction and therefore one force: Because the tube of water has mass, it participates in a long-range, gravitational interaction. Thus, it experiences a gravitational force $\mathbf{F}_g$ (Figure 5.31).

With ρ as the density of water, the tube's mass is ρAh:

$$\underbrace{\text{mass}}_{m} = \underbrace{\text{density}}_{\rho} \times \underbrace{\text{volume}}_{Ah}. \qquad (5.20)$$

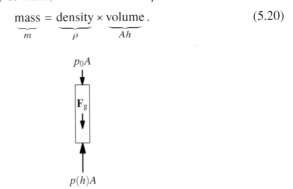

Figure 5.31 The freebody diagram of the tube of water.

Thus, $\mathbf{F}_g$, the third force acting on the tube, has magnitude mg or ρgAh.

The tube of water isn't accelerating, so the three forces must add to zero. The "must add to zero" formulation is a compact alternative to saying, "We know that the forces add to zero." It also avoids the mind-projection fallacy: This use of the model verb "must" is epistemic, reflecting our knowledge or beliefs – for example: "It's late, so he must have reached home by now." (The contrasting modal use is deontic, reflecting external compulsion: "Travelers must show identification at the border.") Recalling an earlier example of this use (Section 5.1): When you stand peacefully on the ground, your acceleration is zero, so the net force on you *must be* zero.

To return from the grammatical interlude, the net force on the tube must similarly be zero (because the tube isn't accelerating). To translate this constraint into an equation, we need a coordinate system. Because the forces are all vertical, we need only one axis – say, the z axis (the usual choice for a vertical axis). We also need a positive z direction – say, downward.

With these choices, the gravitational force (which points downward) and the pressure force on the top surface (due to air pressure and pointing downward) have positive z components; and the pressure force on the bottom surface (due to water pressure and pointing upward) has a negative z component. The result is the long-sought equation:

$$p_0 A - p(h)A + \rho gAh = 0. \qquad (5.21)$$

The area A is common to all terms, so it cancels. Good: It was arbitrary anyway.

$$p_0 - p(h) + \rho gh = 0. \qquad (5.22)$$

Solving for the pressure $p(h)$ as a function of depth h gives the so-called condition for hydrostatic equilibrium:

$$p(h) = p_0 + \rho gh. \qquad (5.23)$$

Thus, the water pressure starts at the surface at atmospheric pressure p_0, increasing linearly with depth at a rate determined by the water's density ρ and gravity's strength g.

With $\rho = 10^3$ kilograms per cubic meter and $g \approx 10$ meters per second squared, the increase in pressure (ρgh) equals the sea-level atmospheric pressure (p_0) when the depth (h) is roughly 10 meters:

$$\underbrace{10^3 \,\mathrm{kg\,m^{-3}}}_{\rho} \times \underbrace{10\,\mathrm{m\,s^{-2}}}_{g} \times \underbrace{10\,\mathrm{m}}_{h} = \underbrace{10^5\,\mathrm{Pa}}_{p_0}. \qquad (5.24)$$

Thus, for every 10 meters of depth, the pressure increases by 1 atmosphere (a pressure unit defined as the sea-level pressure p_0) – which is more than enough for a diver to feel the increase.

The result (5.23) concludes our study of Newton's second law with zero acceleration (constant velocity). Our next step is to include acceleration. Otherwise we cannot explain much motion in our world where bodies usually accelerate. Because acceleration is such a subtle concept, in the next chapter (Chapter 6), you meet acceleration alone – separate from force. In Chapter 7, we reunite acceleration and force to make the full second law.

5.8 Problems

5.1 In finding the normal-force magnitude N in (5.4), I claimed that the perpendicular portion of the gravitational force has magnitude $mg \cos \theta$. In finding the static-friction magnitude f, given in (5.5), I claimed that the parallel portion has magnitude $mg \sin \theta$. Prove these claims by applying trigonometry to a triangle formed by the gravitational force and its parallel and perpendicular portions.

5.2 I sledded down the 30-degree hill of Section 5.5 far faster than I've ever sledded. Redo the analyses using the more realistic slope of 20 degrees, first without and then with dynamic friction. You can approximate the trigonometric functions using $\sin 20° \approx 1/3$ and $\cos 20° \approx 0.9$. Then summarize the no-friction analysis with formulas for F_{contact} and v as functions of θ, m, g, ρ, and A_{cs}.

5.3 Check that the analysis of the hanging mass (Section 5.6.2) has given a reasonable result by analyzing the special case where string 3 hangs vertically – that is, with $\theta_3 = 90°$ (Figure 5.32). What should the three string tensions be (just based on thinking about the system and without calculating)? Do your predictions agree with the calculated string tensions (5.19) when $\theta_3 = 90°$? If not, then your predictions, the calculations leading to (5.19), or both must be wrong – if so, try to find the error.

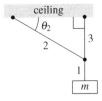

Figure 5.32 The limiting case of Figure 5.23 with $\theta_3 = 90°$.

5.4 Can we hold a tug-of-war (Section 5.6.1) where we move at the same constant velocity, nonzero relative to the ground, yet tug to a draw (perhaps on ice)? If it's not possible, explain why not. If it's possible, draw consistent freebody diagrams showing the forces on you, me, the string, and the earth.

5.5 The tug-of-war analysis of Section 5.6.1 assumed that you and I have the same mass m. Redraw the four freebody diagrams assuming that your mass is m and mine is $2m$.

5.6 Reanalyze the tug-of-war of Section 5.6.1 using a rope instead of a string. (A rope, in physics code, is a string with mass.) Give each of us a mass of 60 kilograms and the rope a mass of 30 kilograms (this rope is unrealistically massive, perhaps by a factor of 3, so that its mass is noticeable on the diagram). Draw consistent freebody diagrams showing the forces on me, you, the rope, and the earth.

5.7 A common trap is thinking that tension T is always mg. This trap is especially tempting in statics. To see that T need not equal mg, look at the pulley arrangement in Figure 5.33. Then find the force F with which you must pull downward in order to raise (or lower) the mass at constant speed.

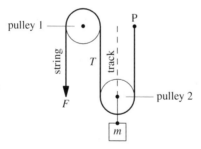

Figure 5.33 A system of pulleys where T isn't mg. The pulleys are massless, pulley 1's axle is fixed to the wall, pulley 2's axle can move vertically along the track, and the string is attached to the wall at point P.

In finding F, you will also find the string tension T. Does $T = mg$? (It shouldn't!) Based on the relation between T and mg, explain the benefit of this pulley arrangement.

5.8 In this problem, you use a simple model of a person standing on the ground in order to calculate and thereby understand the internal forces that we experience every day. Thus, imagine yourself divided into three stacked blocks, each of mass $m/3$ (Figure 5.34).

Figure 5.34 A three-block model of you standing on the ground.

a. Make four freebody diagrams: for each block and for the earth (which includes the ground). The net force on each body should be zero!

b. Connect third-law pairs with dashed lines as in Figure 5.22. Make sure that the two forces in a third-law pair are equal (in magnitude) and opposite (in direction).

5.9 Show that the tension is constant throughout a frictionless string (which is by definition also massless). Do so by assuming that the tension changes and reaching a contradiction. For simplicity, use a straight string. (Although the constant-tension conclusion applies to any string shape, the proof for a general shape requires calculus.)

5.10 The analysis of the equal-mass Atwood machine (Section 5.6.3) omitted an important body: the pulley. Give the pulley mass m_{pulley}, and make its freebody diagram. Express any force magnitudes in terms of m, m_{pulley}, and g. *Hint*: Make the freebody diagram for the string.

5.11 Pretend that the atmosphere is a giant pool of air at constant density (its sea-level density). How deep is the pool? In other words, how high does this atmosphere reach (defined as where the pressure has fallen to zero)? This height H is called the atmosphere's scale height.

5.12 In this problem, you find the tension in a hanging steel cable (Figure 5.35). The cable isn't massless (in physics code, it's a rope). It has length l, cross-sectional area A, and density ρ.

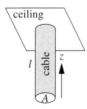

Figure 5.35 A steel cable hanging from the ceiling (Problem 5.12). How does its tension vary with z (the distance above the bottom of the cable)?

a. What's the tension T as a function z?

b. A solid breaks when the *tensile stress* within it exceeds its *breaking strength*. Tensile stress, analogous to pressure in a liquid, is the tension divided by the cross-sectional area. For steel, the breaking strength is roughly 10^9 pascals. How long can the cable be before it breaks simply because of gravity?

5.13 Extend the analysis of constant-velocity bicycling (Section 5.4) to include wind. You still bicycle in the positive x direction, directly east, at constant speed. However, now a wind blows in the positive y direction (directly from the south).

Make a freebody diagram of the bicycle – considered as a composite body that also includes you – showing the three forces acting on the bicycle: $\mathbf{F}_{drag}$, $\mathbf{F}_{contact}$, and $\mathbf{F}_g$. Because of the two-dimensionality of paper, you need to draw two views: a side view showing any xz portions of these forces and a top view showing any xy portions of these forces. To get you started, Figure 5.36 shows those views for windless bicycling (Section 5.4). Modify these diagrams to include the effect of wind.

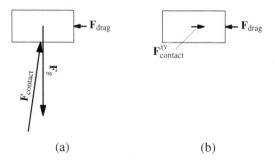

(a) (b)

Figure 5.36 Two views of the three-dimensional freebody diagram for the windless bicycling of Section 5.4. (a) The side view, showing the forces' xz portions. (b) The top view, showing the forces' xy portions. In this view, the contact force gets an xy superscript to indicate that not all of the force is visible.

5.14 Figure 5.37a shows an incomplete freebody diagram of a plane flying to the right at constant altitude and constant velocity (thus, in a horizontal line at constant speed).

a. Complete the freebody diagram by including $\mathbf{F}_{\text{air on plane}}$. If the plane, still moving in a straight line at constant speed, were climbing at a steady rate, how would your diagram change?

b. This force is composed of three more-familiar portions: thrust, lift, and drag. Draw a vector sum of thrust, lift, and drag showing them adding

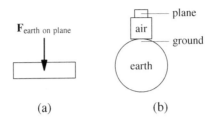

Figure 5.37 (a) An incomplete freebody diagram of a plane flying at constant altitude and to the right at constant velocity. (b) The system of the plane, the air, and the earth (the ground is part of the earth).

up to $\mathbf{F}_{\text{air on plane}}$. (The diagram is easier to make if you draw them in the order thrust, lift, and then drag.)

c. Make labeled freebody diagrams of the air and of the earth, which includes the ground (Figure 5.37b). For ease of drawing, chose the air to be the air in an imaginary box extending vertically from just below the plane to the ground and extending horizontally 10 or 20 meters in each direction – to make the air's mass one-half of the plane's mass. Use arrow lengths to indicate rough relative force magnitudes.

d. Use dashed paths to connect third-law pairs across diagrams. Ensure that every force belongs to exactly one third-law pair, and label each dashed path with the two interacting bodies (for example, earth–air).

5.15 (This problem uses calculus.) In the analysis of the pressure in a lake (Section 5.7), we implicitly assumed that the fluid's density ρ was constant. For water, which is highly incompressible, this assumption is usually fine (it gets dodgy only toward the bottom of the oceans). For air, however, this assumption fails in an important situation: the atmosphere. In this problem, you relax the constant-density assumption of Problem 5.11 and find how atmospheric pressure and density vary with height above sea level – assuming that the temperature does not vary from its value at sea level. This isothermal assumption is another approximation but much more accurate than the constant-density assumption.

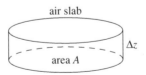

Figure 5.38 A cylindrical slab of air with height Δz and cross-sectional area A.

a. Find the pressure change Δp over the height of a thin cylindrical slab of air (Figure 5.38) with height Δz and cross-sectional area A. The pressure change is defined as

$$\Delta p \equiv p(z + \Delta z) - p(z),\qquad(5.25)$$

where $p(z)$ is the pressure at height z (sea level is $z = 0$).

You'll need the relation between density ρ and pressure, which comes from the ideal-gas law suitably transformed:

$$\rho = \frac{m_{\text{molar}}}{RT}p,\qquad(5.26)$$

where m_{molar} is the molar mass of air (roughly 30 grams per mole), R is the universal gas constant (roughly 8 joules per mole per Kelvin), and T is the absolute or Kelvin (not Celsius!) temperature of the air (roughly 300 Kelvin). (For air at sea-level temperature, the ideal-gas law (5.26) becomes

$$\rho \approx 1.2\,\text{kg}\,\text{m}^{-3} \times \frac{p}{10^5\,\text{Pa}}.\qquad(5.27)$$

As a check: At sea level, where $p \approx 10^5$ pascals, the formula correctly gives $\rho \approx 1.2$ kilograms per cubic meter.)

b. Use the thin-slab approximation,

$$\Delta p \approx \frac{dp}{dz}\Delta z,\qquad(5.28)$$

to find the differential equation for p (that is, to find dp/dz).

c. Solve the equation to find the pressure p as a function of p_0, z, m_{molar}, R, T, and g.

d. Use the ideal-gas law (5.26) to find the density $\rho(z)$ as a function of z, m_{molar}, R, T, g, and ρ_0 (the sea-level density).

e. At airplane cruising altitude, roughly 33,000 feet or 10 kilometers, what's the density relative to the density at sea level (ρ/ρ_0)? At the atmosphere's scale height H, as computed in Problem 5.11, what is the density relative to the density at sea level?

6

Describing Changing Motion: Acceleration

Newton's laws don't care about motion or velocity directly. For as you saw in Section 3.4, you can make a body have any velocity that you like merely by using a new inertial reference frame. Thus, Newton's second law proclaims: "Force *changes* motion!"

In just three words, it unites two subtle ideas: force, from the physical world, and changing motion, from the mathematical world. We studied force – the physical quantity that formalizes our intuitive notion of a push or a pull – in Chapters 1, 2, and 5. In this chapter, we study acceleration: the mathematical quantity that describes changing motion. In Chapter 7, we unite physics and mathematics and use the second law in its full glory.

Understanding acceleration requires distinguishing a vector quantity from its related scalars and distinguishing a quantity's average value from its instantaneous value. These distinctions are introduced using a familiar quantity, velocity (Section 6.1), the doorway to acceleration. After you learn acceleration's formal definition (Section 6.2), you befriend acceleration by calculating it for motion of ever-increasing generality: constant-speed motion around a circle (Section 6.3), constant-speed motion around a noncircular path (Section 6.4), and varying-speed motion around a circular path (Section 6.5). Finally, the key points from the preceding calculations are combined to describe acceleration in the general case: varying-speed motion along a general path (Section 6.6).

6.1 Velocity

A body's motion is most directly described by its velocity – for which, unlike for acceleration, our intuitions are sound. Thus, I use velocity to introduce the two ideas essential to understanding acceleration: the distinction between a vector and the scalars derived from it (Section 6.1.1) and the distinction between the instantaneous and the average value of a quantity (Section 6.1.2).

6.1.1 Vectors and the Scalars That They Make

We already know intuitively that velocity is a vector: a quantity with magnitude and direction. Thus, we might describe a bird's velocity as

$$\mathbf{v} = \underbrace{40 \, \text{kph}}_{\text{magnitude}} \; \underbrace{\text{northwest}}_{\text{direction}}. \tag{6.1}$$

From a vector, we can make related scalars: related quantities without direction. The most important such scalar is the vector's magnitude. The magnitude of velocity is so important that it has a special, familiar name: the speed. Magnitude is notated either in italics (v) or by absolute-value bars around the vector ($|\mathbf{v}|$).

Many misconceptions about acceleration arise from confusing speed with velocity. This confusion probably arises from how we learn scientific terms. When we first meet a scientific term, it gets attached to our closest ordinary-language concept. Although speed is an ordinary-language concept, velocity is not: No ordinary-language concept about motion contains both magnitude and direction (at least in English). So when we hear "velocity," we attach it to speed. This confusion is then reproduced even in scientific texts: I often read that "the *velocity* of light is 3×10^8 meters per second," when its speed is meant.

The remedy is always to distinguish between velocity and speed. Use velocity if you mean the vector; use speed if you mean its magnitude. Thus, never say, "The bird's velocity is increasing (or decreasing)," or, "Its velocity is positive." Velocity, being a vector, has no sign and so cannot increase or decrease.

The other derived scalars are a vector's components: its amounts in various directions. For a train moving to the left (the direction) at 25 meters per second (the magnitude), with the $+x$ direction pointing to the right:

$$\begin{aligned} v_x &= -25 \, \text{m s}^{-1}; \\ v_y &= 0. \end{aligned} \tag{6.2}$$

The direction subscript, x or y here, not only tells us which component is intended, it also distinguishes a component (an italic letter with a direction subscript) from a magnitude (an italic letter with no direction subscript).

6.1.2 Instantaneous versus Average Velocity

Velocity is short for *instantaneous* velocity – meaning a body's velocity at a particular moment. The modifier "instantaneous" implies that the velocity can change: The bird that was flying 40 kilometers per hour northwest is right now flying 30 kilometers per hour southwest.

As immigrants know, change is complicated. With quantities, a powerful way to tame change is to work with the quantity's average value. This idea is essential in understanding acceleration, which describes changing velocity. Velocity itself describes how position changes, so it already contains the idea of change. It's also simpler than acceleration. Thus, let's start with average velocity $\mathbf{v}_{avg}$. The average is defined over a time interval $t_1 \ldots t_2$:

$$\text{average velocity } \mathbf{v}_{avg} \equiv \frac{\mathbf{r}_2 - \mathbf{r}_1}{t_2 - t_1}, \tag{6.3}$$

where $\mathbf{r}_1$ and $\mathbf{r}_2$ are the body's positions (vectors!) at t_1 and t_2, respectively. (As I mentioned after (1.24), the triple-equals sign, $\equiv$, means "equality by definition" or "is defined to be.")

In (6.3), the denominator is the interval's duration; as a difference of scalars (the two times), it's a scalar. The numerator, as a difference of vectors (the two positions), is a vector. It's the body's change in position, also called its *displacement*. In words,

$$\text{average velocity} \equiv \frac{\text{displacement}}{\text{duration}}. \tag{6.4}$$

The duration is often written more compactly as Δt. The delta symbol, Δ, a notational contribution of Leibniz, means "change in." Thus, Δt means the change in t: the duration. In particular, Δ means the new value minus the old value (not the old value minus the new value). Thus,

$$\Delta t \equiv t_2 - t_1 \text{ (not } t_1 - t_2!). \tag{6.5}$$

Similarly, the displacement, which is the vector *from* $\mathbf{r}_1$ *to* $\mathbf{r}_2$, is often written as $\Delta \mathbf{r}$ (the change in $\mathbf{r}$):

$$\Delta \mathbf{r} \equiv \mathbf{r}_2 - \mathbf{r}_1 \text{ (not } \mathbf{r}_1 - \mathbf{r}_2). \tag{6.6}$$

Then, in compact form, (6.3) and (6.4) become

$$\mathbf{v}_{avg} \equiv \frac{\Delta \mathbf{r}}{\Delta t}. \tag{6.7}$$

Here is an average-velocity calculation about the earth's orbit (Figure 6.1a).

▶ *What's the earth's average velocity (and related scalars) over one-half of its annual orbit around the sun?*

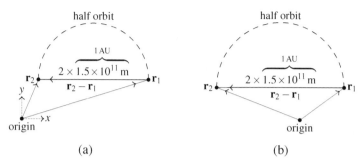

Figure 6.1 One-half of the earth's orbit. The origin of the coordinate system can be located anywhere with no change in the velocities. (a) One possible origin. (b) A second possible origin. No matter where the origin lies, the displacement vector, $\mathbf{r}_2 - \mathbf{r}_1$, is always $2 \times 1.5 \times 10^{11}$ meters to the left. (For clarity, the vectors $\mathbf{r}_1$, $\mathbf{r}_2$, and $\mathbf{r}_2 - \mathbf{r}_1$ have their direction-indicating arrows slightly behind their tips.)

The orbital radius is 1 astronomical unit (AU) or 1.5×10^{11} meters. The earth's displacement – the vector from $\mathbf{r}_1$ to $\mathbf{r}_2$ – has magnitude equal to the orbital diameter and points to the left (don't forget the direction!). The duration is one-half of a year. From (6.3),

$$\mathbf{v}_{\mathrm{avg}} \approx \frac{\overbrace{2 \times 1.5 \times 10^{11} \text{ m to the left}}^{\mathbf{r}_2-\mathbf{r}_1}}{\underbrace{\frac{1}{2} \times 3 \times 10^7 \text{ s}}_{\approx 1 \text{ yr}}} = \underbrace{2 \times 10^4 \text{ m s}^{-1}}_{\text{magnitude}} \underbrace{\text{to the left}}_{\text{direction}}. \quad (6.8)$$

The individual positions, $\mathbf{r}_1$ and $\mathbf{r}_2$, depend on the reference frame's origin (because each is a vector from the origin to the earth's starting or ending position). However, the displacement vector, as their difference $\mathbf{r}_2 - \mathbf{r}_1$, doesn't depend on the location of the origin: Even in a new frame, the displacement is 3×10^{11} meters to the left (Figure 6.1b). Thus, average velocity doesn't depend on the location of the origin. (This conclusion is valid as long as the new reference frame isn't moving relative to the old reference frame – an important point that reappears in Section 6.2.1 when we discuss average acceleration and inertial reference frames.)

For the earth's average velocity, the first related scalar is its magnitude:

$$|\mathbf{v}_{\mathrm{avg}}| = 20 \text{ km s}^{-1}. \quad (6.9)$$

The other related scalars are the components:

$$v_x^{\mathrm{avg}} = -20 \text{ km s}^{-1};$$
$$v_y^{\mathrm{avg}} = 0. \quad (6.10)$$

(The v_x component is negative because $\mathbf{v}_{\mathrm{avg}}$ points in the $-x$ direction.)

6.1.3 Average Speed

So far, you have met average velocity,

$$\mathbf{v}_{\text{avg}} = \frac{displacement \text{ (a vector) from } \mathbf{r}_1 \text{ to } \mathbf{r}_2}{\text{duration}}, \qquad (6.11)$$

and its magnitude,

$$|\mathbf{v}_{\text{avg}}| = \frac{distance \text{ (a scalar) from } \mathbf{r}_1 \text{ to } \mathbf{r}_2}{\text{duration}}. \qquad (6.12)$$

The remaining velocity-related quantity is average speed v_{avg} (a scalar). Surprisingly, it's not equal to $|\mathbf{v}_{\text{avg}}|$. Rather, its numerator, though also a scalar, is slightly different than $|\mathbf{v}_{\text{avg}}|$'s:

$$v_{\text{avg}} = \frac{distance\ covered \text{ (a scalar) traveling from } \mathbf{r}_1 \text{ to } \mathbf{r}_2}{\text{duration}}. \qquad (6.13)$$

To see the distinction, look again at the earth's half-orbit (Figure 6.1a). Whereas the distance from $\mathbf{r}_1$ to $\mathbf{r}_2$ is the orbital diameter, the distance *covered* as the earth travels from $\mathbf{r}_1$ to $\mathbf{r}_2$ is one-half of the orbital circumference:

$$v_{\text{avg}} \approx \frac{\frac{1}{2} \times \overbrace{2\pi \times 1.5 \times 10^{11}\,\text{m}}^{\text{circumference}}}{\frac{1}{2} \times \underbrace{\pi \times 10^7\,\text{s}}_{\approx 1\,\text{yr}}} = 30\,\text{km s}^{-1}. \qquad (6.14)$$

The average speed is larger than $|\mathbf{v}_{\text{avg}}|$, which, from (6.9), is 20 kilometers per second. The general difference between the two quantities lies in the order of applying the two operations that each contains. For $|\mathbf{v}_{\text{avg}}|$, we first calculate the vector average of $\mathbf{v}$, using (6.3) or (6.7), and only then do we take the magnitude to get a scalar. For v_{avg}, we first take the magnitude of $\mathbf{v}$ to get a scalar, and only then do we average this scalar over time; this difficult direct calculation turns out to be equivalent to the simpler (6.13).

With this background, you can solve the following problem developed by J. W. Warren [25, p. 2], who taught physics in London at Brunel College (now Brunel University) and to whom this book is dedicated. He reported that correct answers to these questions are "practically never obtained." You know enough to provide a counterexample, as long as you reason from first principles!

Problem. A particle moves in the path shown [in Figure 6.2], the speed increasing uniformly with time in the semicircular section, from 10 meters per second to 12 meters per second. For this section of the path, calculate the averages of the velocity and the acceleration. (Use $\pi = 22/7$.)

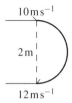

Figure 6.2 A particle moving along a semicircular track. The particle starts the semicircle moving at 10 meters per second and leaves it moving at 12 meters per second (with the speed increasing uniformly with time).

The average acceleration will be calculated in Section 6.2.1. For now, using the definition (6.7), let's calculate the average velocity. The displacement $\Delta\mathbf{r}$ (the change in the particle's position) is the vertical diameter of the circle and points from top to bottom (from the starting to the ending position):

$$\Delta\mathbf{r} = 2 \text{ m downward.} \tag{6.15}$$

To find the duration Δt, turn around the definition of average speed (6.13):

$$\Delta t = \frac{\text{distance covered traveling along the semicircle}}{v_{\text{avg}}}. \tag{6.16}$$

With the provided approximation $\pi = 22/7$, the numerator becomes

$$\text{distance covered} = \underbrace{\frac{1}{2} \times \pi \times 2 \text{ m}}_{\text{circumference}} \approx \frac{22}{7} \text{ m.} \tag{6.17}$$

Over the semicircular section, the particle's speed increases uniformly with time (meaning steadily) from 10 to 12 meters per second, so v_{avg} is the midpoint:

$$v_{\text{avg}} = 11 \text{ m s}^{-1}. \tag{6.18}$$

Now put this result and the distance covered (6.17) into the duration (6.16):

$$\Delta t \approx \frac{\overbrace{(22/7) \text{ m}}^{\text{distance covered}}}{\underbrace{11 \text{ m s}^{-1}}_{v_{\text{avg}}}} = \frac{2}{7} \text{ s.} \tag{6.19}$$

Finally, combine this duration with the displacement $\Delta\mathbf{r}$ from (6.15):

$$\mathbf{v}_{\text{avg}} \approx \frac{\overbrace{2 \text{ m downward}}^{\Delta\mathbf{r}}}{\underbrace{(2/7) \text{ s}}_{\approx \Delta t}} = 7 \text{ m s}^{-1} \text{ downward.} \tag{6.20}$$

6.2 Acceleration Defined

If I happen to drop a wet, slippery drinking glass, my eyes could probably judge that, just before impact, the glass was moving downward at 10 miles per hour (about 4 meters per second). However, the same eyes would give me only a vague idea of how the glass's velocity was changing – of the glass's acceleration. Seeing acceleration requires looking with the eyes of mathematics: We need to define acceleration and only thereafter to make the definition intuitive. The definition begins with average acceleration.

6.2.1 Average Acceleration

Average acceleration, analogously to average velocity, is defined for a time interval $t_1 \ldots t_2$:

$$\text{average acceleration } \mathbf{a}_{\text{avg}} \equiv \frac{\text{change in velocity}}{\text{duration}} = \frac{\mathbf{v}_2 - \mathbf{v}_1}{t_2 - t_1}. \quad (6.21)$$

At each of these times t_1 and t_2, we determine the body's velocity (a vector!), getting $\mathbf{v}_1$ and $\mathbf{v}_2$, respectively (analogously to getting $\mathbf{r}_1$ and $\mathbf{r}_2$ in calculating average velocity). Their difference $\mathbf{v}_2 - \mathbf{v}_1$ is abbreviated $\Delta\mathbf{v}$. Compactly and analogously to (6.7) for average velocity,

$$\mathbf{a}_{\text{avg}} \equiv \frac{\Delta\mathbf{v}}{\Delta t}. \quad (6.22)$$

As an example, my daughters and I wait at a crosswalk, and a car approaches us at 108 kilometers per hour from the left. At $t = 0$, the driver sees us and, at $t = 2$ seconds, starts braking. At $t = 8$ seconds, the car stops.

▶ *What's the car's average acceleration (and related scalars) during braking?*

Beware of sign errors! But if you follow the average-acceleration recipe (6.21), keeping track of what time belongs with what velocity, you cannot go wrong. The braking interval starts at $t_1 = 2$ seconds, when the driver has finished reacting, and ends at $t_2 = 8$ seconds, when the car stops. At these times,

$$\mathbf{v}_1 = 108 \text{ kph to the right, and} \\ \mathbf{v}_2 = 0. \quad (6.23)$$

Being careful about the minus signs and about the order of $\mathbf{v}_1$ and $\mathbf{v}_2$ in the numerator and of t_1 and t_2 in the denominator,

$$\mathbf{a}_{\text{avg}} \equiv \frac{\Delta\mathbf{v}}{\Delta t} = \frac{\overset{\mathbf{v}_2}{\overbrace{0} - \overset{\mathbf{v}_1}{\overbrace{108 \text{ kph to the right}}}}}{\underset{t_2}{\underbrace{8\,\text{s}}} - \underset{t_1}{\underbrace{2\,\text{s}}}}$$

$$= \frac{-108 \text{ kph to the right}}{6\,\text{s}}$$

$$= -18 \text{ kph s}^{-1} \text{ to the right.} \quad (6.24)$$

I try to avoid minus signs because each one invites a sign error. Here, avoiding the minus sign in -18 is possible by specifying the direction as "to the left":

$$\mathbf{a}_{\text{avg}} = +18\,\text{kph s}^{-1} \text{ to the left.} \tag{6.25}$$

The average acceleration points *opposite* to the velocity!

From $\mathbf{a}_{\text{avg}}$, we can compute the related scalars. Its magnitude, whether from (6.24) or (6.25), is a positive scalar:

$$|\mathbf{a}_{\text{avg}}| = +18\,\text{kph s}^{-1}. \tag{6.26}$$

Its only nonzero component is its x component, a *negative* scalar:

$$a_x^{\text{avg}} = -18\,\text{kph s}^{-1}. \tag{6.27}$$

Although a magnitude cannot be negative, a component can.

One of the great difficulties in mastering Newton's laws is distinguishing velocity from acceleration – in order to avoid the ancient "force causes velocity" trap. In one dimension, however, velocity and acceleration are particularly easy to confuse because they always lie on the same line (even if they point in opposite directions). In two (or more) dimensions, the distinction is easier to make. Thus, let's return to the particle of Section 6.1.3 on the semicircular track.

▶ *Over the semicircular section, what's the particle's average acceleration?*

In $\mathbf{a}_{\text{avg}}$'s definition (6.22), we have already calculated the duration Δt, given in (6.19), and need just to calculate the velocity change $\Delta \mathbf{v}$. The starting and ending velocities are, respectively,

$$\begin{aligned} \mathbf{v}_1 &= 10\,\text{m s}^{-1} \text{ to the right, and} \\ \mathbf{v}_2 &= 12\,\text{m s}^{-1} \text{ to the left.} \end{aligned} \tag{6.28}$$

Their difference $\Delta \mathbf{v}$ is $\mathbf{v}_2 - \mathbf{v}_1$ (not $\mathbf{v}_1 - \mathbf{v}_2$):

$$\begin{aligned} \Delta \mathbf{v} &= \underbrace{12\,\text{m s}^{-1} \text{ to the left}}_{\mathbf{v}_2} - \underbrace{10\,\text{m s}^{-1} \text{ to the right}}_{\mathbf{v}_1} \\ &= (12\,\text{m s}^{-1} \text{ to the left}) - (-10\,\text{m s}^{-1} \text{ to the left}) \\ &= 22\,\text{m s}^{-1} \text{ to the left.} \end{aligned} \tag{6.29}$$

This calculation avoided several common traps: forgetting that the numerator is $\Delta \mathbf{v}$ (the change in velocity, a vector) rather than Δv (the change in speed); forgetting to use the same reference direction, to the left, for both velocities (a species of sign error); and forgetting the minus sign when converting 10 meters per second to the right into -10 meters per second to the left.

With $\Delta \mathbf{v}$ from (6.29) and with the approximate duration from (6.19),

$$\mathbf{a}_{\text{avg}} \approx \frac{\overbrace{22\,\text{m s}^{-1} \text{ to the left}}^{\Delta \mathbf{v}}}{\underbrace{(2/7)\,\text{s}}_{\approx \Delta t}} = 77\,\text{m s}^{-2} \text{ to the left.} \tag{6.30}$$

In magnitude, this acceleration is almost eight times as large as the gravitational acceleration! If you ride on the particle, you feel unpleasantly large forces as you round the semicircle.

In the next section (Section 6.2.2), we look at acceleration's crazy units – for example, kilometers per hour per second in (6.24) or meters per second squared in (6.30). Before we do so, however, I mention an important characteristic of average acceleration: Unlike average velocity, average acceleration is the same in all inertial frames. This characteristic leads to another distinction between velocity and acceleration and a reminder that force is connected to acceleration.

To see why average acceleration is the same in all inertial frames, look back at its definition (6.21), which requires first that we compute a body's starting velocity $\mathbf{v}_1$ and ending velocity $\mathbf{v}_2$. In a new, moving (and nonrotating) frame, $\mathbf{v}_1$ and $\mathbf{v}_2$ are different from their values in the old frame. From (3.5),

$$\begin{aligned} \mathbf{v}_1^{\text{new}} &= \mathbf{v}_1^{\text{old}} - \mathbf{v}_{\text{frame}}(t_1), \\ \mathbf{v}_2^{\text{new}} &= \mathbf{v}_2^{\text{old}} - \mathbf{v}_{\text{frame}}(t_2), \end{aligned} \tag{6.31}$$

where $\mathbf{v}_{\text{frame}}(t_1)$ and $\mathbf{v}_{\text{frame}}(t_2)$ are the new frame's velocities (relative to the old frame) when the clock starts and stops, respectively.

However, if the new frame is inertial, it moves with constant velocity relative to the old frame. That is, $\mathbf{v}_{\text{frame}}$ is constant. So, the change in velocity $\mathbf{v}_2 - \mathbf{v}_1$, the numerator of (6.21), is the same in both frames!

$$\underbrace{(\mathbf{v}_2^{\text{old}} - \mathbf{v}_{\text{frame}})}_{\mathbf{v}_2^{\text{new}}} - \underbrace{(\mathbf{v}_1^{\text{old}} - \mathbf{v}_{\text{frame}})}_{\mathbf{v}_1^{\text{new}}} = \mathbf{v}_2^{\text{old}} - \mathbf{v}_1^{\text{old}}. \tag{6.32}$$

The duration $t_2 - t_1$, the denominator of (6.21), is also the same in both frames: In Newtonian mechanics, all reference frames share the same universal clock (universal, or absolute, time goes away in relativistic mechanics, which is touched upon in Section 8.3.2). Thus, the body's (average) acceleration is the same in both frames – unlike its average velocity.

Therefore, for Newton's second law to be valid in all inertial frames – the frames tested and defined by the first law (Section 3.2) – the second law cannot mention velocity and is almost forced to use acceleration. Once again, force is connected to acceleration.

6.2.2 Acceleration's Crazy Units

Acceleration (the vector), its magnitude, and its components have crazy units. Their story begins with a concept more fundamental than units – namely, dimensions (length, mass, time, etc.). The dimensions of acceleration, notated [acceleration], come from the dimensions of velocity:

$$[\text{acceleration}] = \frac{\overbrace{[\text{velocity}]}^{\text{from } \Delta v \text{ in (6.22)}}}{\underbrace{\text{time}}_{\text{from } \Delta t \text{ in (6.22)}}}. \tag{6.33}$$

The dimensions of velocity are length per time, so (6.33) becomes

$$[\text{acceleration}] = \frac{\text{length per time}}{\text{time}}. \tag{6.34}$$

Because "per" means "divided by,"

$$[\text{acceleration}] = \frac{\text{length}}{(\text{time})^2}. \tag{6.35}$$

In other words, acceleration has dimensions of length per time squared. The "time squared" foreshadows the unit craziness to come.

Stating an actual acceleration requires a unit system: an agreement on how to measure acceleration's various dimensions. Based on (6.34), units must be chosen for length, for time in the numerator (in the "per time"), and for time in the denominator.

For example, you could chose kilometers, hours, and seconds, respectively, as in the average acceleration calculated for the braking car in (6.25) and restated in expanded form here:

$$\mathbf{a}_{\text{avg}} = (18 \ \underbrace{kilometers}_{\text{length}} \text{ per } \underbrace{hour}_{\text{numerator time}} \text{ to the left) per } \underbrace{second}_{\text{denominator time}}. \tag{6.36}$$

Or you might choose miles, hours, and seconds, respectively. With 1 mile approximately 1.6 kilometers, 18 kilometers is roughly 11 miles. Thus, for the braking car, (6.25) becomes, in the expanded style of (6.36),

$$\mathbf{a}_{\text{avg}} \approx (11 \text{ miles per hour to the left) per second.} \tag{6.37}$$

The unusual parentheses in (6.36) and (6.37) clarify the units' meaning. Inside the parentheses is the change in velocity, which is itself a rate and therefore contains an inverse time (here, "per hour"). Outside the parentheses, the "per second" says how long the change described in the parentheses requires. Thus, every second, the car's velocity changes, on average, by approximately 11 miles per hour to the left. Most textbooks don't include these parentheses. However, their explanatory value makes them worth including. In reading to myself, I indicate them with a pause: "11 miles per hour to the left... per second."

The two time units can even be the same, as the following example illustrates.

▷ *What's the car's average x acceleration during braking, in SI (metric) units?*

The SI or metric unit of length is the meter; the SI unit of time is the second. Thus, the velocity units of kilometers per hour are not metric. To convert them to metric, convert kilometers to meters and hours to seconds:

$$a_x^{\text{avg}} = \frac{-18 \ \cancel{\text{km}}}{\cancel{\text{hr}}, \text{s}} \times \underbrace{\frac{1000 \ \text{m}}{1 \ \cancel{\text{km}}}}_{\to \text{meters}} \times \underbrace{\frac{1 \ \cancel{\text{hr}}}{3600 \ \text{s}}}_{\to \text{seconds}} = \frac{-18 \times 1000 \ \text{m}}{3600 \ \text{s}^2}. \qquad (6.38)$$

Numerically, $-18 \times 1000/3600$ is -5. So,

$$a_x^{\text{avg}} = \frac{-5 \ \text{m}}{\text{s}^2}. \qquad (6.39)$$

Using negative exponents is more compact:

$$a_x^{\text{avg}} = -5 \ \text{m} \, \text{s}^{-2}. \qquad (6.40)$$

Written out, a_x^{avg} is -5 meters per second squared. Fortunately, you now know what those bizarre units mean – units that confused me for many years as I puzzled over the square of a second. They mean (meters per second) per second, and a_x^{avg} is (-5 meters per second) per second.

Interestingly, g, the acceleration (magnitude) due to gravity is (10 meters per second) per second. Thus, in magnitude, the car's average acceleration during braking is $0.5g$, which implies that large horizontal forces act on the car's occupants, forces roughly half as strong (on average and in magnitude) as the gravitational forces. In a car crash, the car's acceleration is much greater than g. It's much healthier for the car's occupants if the required forces come from a slightly flexible seat belt than from a hard windshield or steering wheel.

6.2.3 Instantaneous Acceleration

The second and final step in calculating acceleration is to turn average acceleration into instantaneous acceleration. Whereas average acceleration is calculated over a time interval $t_1 \ldots t_2$, instantaneous acceleration is calculated at a single time t. It's the *limit* of the average acceleration as t_1 and t_2 approach t. In this limit, the duration Δt approaches zero and is called dt. Meanwhile, the velocity difference $\Delta \mathbf{v}$ also approaches zero and is written $d\mathbf{v}$:

$$\mathbf{a} \equiv \lim_{\Delta t \to 0} \underbrace{\frac{\Delta \mathbf{v}}{\Delta t}}_{\mathbf{a}_{\text{avg}}} \equiv \frac{d\mathbf{v}}{dt}. \qquad (6.41)$$

Acceleration is the rate of change of velocity with respect to time.

In the simplest case of this definition, a body's velocity changes at a steady rate – for example, steadily adding 3 meters per second downward every second. The average acceleration is then the same over any time interval, including an infinitesimally short one. Thus, the average and instantaneous accelerations are identical. (This case is illustrated in Problem 6.2.)

In the more general case where the velocity does not change steadily, the calculation looks impossible because it ends up in zero (from $\Delta \mathbf{v}$) divided by zero (from Δt). To resolve that problem, we decrease the duration of the interval until it's infinitesimally short, calculating the ratio $\Delta \mathbf{v}/\Delta t$ along the way and seeing what it becomes. This limiting process will be illustrated using the most important motion in more than one dimension: a body moving in a circle at constant speed (Section 6.3). Its analysis was one of Newton's great mathematical achievements as he developed what we now call Newtonian mechanics.

6.3 Circular Motion at Constant Speed

The problem of the semicircular track (Section 6.1.3) causes so much trouble partly because the particle does not move in a straight line. In straight-line motion, as I mentioned on p. 97, the velocity and acceleration always lie along a common line. Thus, even without realizing that force causes acceleration – that it *changes* velocity (rather than producing velocity itself) – you can end up by luck with correct forces and freebody diagrams.

However, because most particles move in curved paths, this luck soon runs out. The canonical example, which even describes most planets' motion reasonably accurately, is motion in a circle.

Circular motion is so difficult, and was so even for the most accomplished mathematician of the 17th century (perhaps of all time), because acceleration and velocity never lie along the same line. Their misalignment exposes our deepest confusion about the second law, the mistaken, pre-Newtonian view held for most of hominid history (and discussed in Section 4.1) that force causes velocity – in contrast to the Newtonian view that force causes acceleration (that force *changes* velocity). On the pre-Newtonian view, an explanation is needed for the body's velocity; and the temptation is to invent a force along the direction of motion, often named $\mathbf{F}_{\text{motion}}$ or $\mathbf{F}_{\text{engine}}$ (as in Figure 5.4b).

To avoid these problems, first remind yourself that force causes acceleration. Second, compute acceleration free from the misconception, over-generalized from one-dimensional motion, that it must lie on the same line as the velocity.

We'll therefore compute the (instantaneous) acceleration from first principles, using the definition of acceleration as the limit of the average acceleration as the time interval goes to zero. Thus, we compute the average acceleration over ever-shorter times. To keep the mathematics from obscuring the new ideas, this analysis will be for a body moving in a circle at constant speed. In Section 6.4, you learn what happens to the acceleration when the path isn't circular (but the speed is constant). In Section 6.5, you learn what happens when the speed varies (but the path is circular). And in Section 6.6, you learn what happens in general.

▶ *Why did I say "at constant speed" rather than "at constant velocity"?*

That question was just to keep you on your toes: Circular motion at constant velocity is impossible because constant velocity means constant in magnitude and direction, so it requires motion in a straight line.

With that reminder of the difference between velocity and its magnitude (speed), imagine a runner moving at constant speed v around a circular track of radius R (Figure 6.3). The runner's velocity points straight ahead of the runner, tangent to the circle, and its magnitude is the constant speed v.

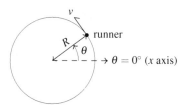

Figure 6.3 A runner moving at constant speed around a circular track. Even though the speed doesn't change, the velocity does change – by changing its direction.

Now we are ready to calculate the runner's acceleration by calculating the runner's average acceleration over ever-shorter time intervals. We shrink the intervals until the result of the average-acceleration calculation becomes clear for an infinitesimally short time interval. That result is the instantaneous acceleration.

Thus, we start with a large Δt, one where the average acceleration is easy to calculate. The calculation requires a coordinate system, in order to specify the runner's position, velocity, and acceleration. For analyzing circular motion, the best system is polar coordinates. As seen from above, the 3-o'clock position, on an imagined overlaid analog clock, is $\theta = 0°$ and also the x axis (following the usual convention, θ increases counterclockwise). The angle θ along with the runner's distance from the center, which here is always R, specify the runner's position. To spare us an unnecessary minus sign, the runner runs counterclockwise – which makes θ increase with time.

▶ *What's the runner's average acceleration over the half-revolution from the 6 o'clock position to the 12 o'clock position (Figure 6.4a)?*

The average acceleration is, from (6.22), the change in velocity divided by the duration. This calculation will be slightly simpler than the calculation of average acceleration in Section 6.2.1 for motion around a semicircular track – slightly simpler because here the moving object's speed is constant.

The runner's starting velocity, at the 6 o'clock position, is v to the right. The ending velocity, at the 12 o'clock position, is v to the left (Figure 6.4b). For $\Delta\mathbf{v}$:

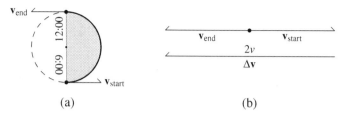

(a)   (b)

Figure 6.4 Calculating $\Delta\mathbf{v}$ for the half-revolution journey. (a) The runner's starting and ending velocities. (b) $\Delta\mathbf{v}$ as their vector difference $\mathbf{v}_{end} - \mathbf{v}_{start}$. It's calculated tip-to-tip: from the tip of $\mathbf{v}_{start}$ to the tip of $\mathbf{v}_{end}$. (I moved the $\Delta\mathbf{v}$ drawing downward to avoid it overlapping the velocity vectors.)

$$\Delta\mathbf{v} = \underbrace{v \text{ to the left}}_{\mathbf{v}_{end}} - \underbrace{v \text{ to the right}}_{\mathbf{v}_{start}} = 2v \text{ to the left.} \tag{6.42}$$

This calculation can easily go wrong in two ways, both leading to the (wrong) conclusion that $\Delta\mathbf{v}$ is zero. To avoid the computational trap, be careful with the minus signs. The starting velocity of v to the right is also $-v$ to the left. Thus,

$$\Delta\mathbf{v} = v \text{ to the left} - (-v \text{ to the left}) = 2v \text{ to the left.} \tag{6.43}$$

To avoid the conceptual trap, be careful with vectors versus scalars. The boldface $\mathbf{v}$ in $\Delta\mathbf{v}$ means that $\Delta\mathbf{v}$ is the result of subtracting vectors. Although the speed v is constant, the velocity $\mathbf{v}$ is not. Thus, although Δv is zero, $\Delta\mathbf{v}$ is not.

The duration Δt is one-half of a full period T. The full period is

$$T = \frac{\text{circumference}}{\text{speed}} = \frac{2\pi R}{v}, \tag{6.44}$$

making the duration

$$\Delta t = \frac{1}{2}T = \frac{\pi R}{v}. \tag{6.45}$$

Putting this Δt and the $\Delta\mathbf{v}$ from (6.42) into the average-acceleration definition (6.22) gives the answer to the triangle question:

$$\mathbf{a}_{avg} \equiv \frac{\Delta\mathbf{v}}{\Delta t} = \frac{2v \text{ to the left}}{\pi R/v} = \frac{2}{\pi}\frac{v^2}{R} \text{ to the left.} \tag{6.46}$$

This average acceleration approximates the instantaneous acceleration best at the midpoint of the time interval, when the runner is at 3 o'clock ($\theta = 0$).

To improve the approximation, we just recalculate $\mathbf{a}_{avg}$ using a shorter interval, choosing Δt to be only one-quarter of a period (Figure 6.5a). One period is $2\pi R/v$, so the shorter time interval is

$$\Delta t = \frac{1}{4}\frac{2\pi R}{v} = \frac{\pi R}{2v}. \tag{6.47}$$

To make this average acceleration comparable to the preceding result (6.46) for one-half of a period, we put this journey's midpoint also at the 3-o'clock position (at $\theta = 0°$). The journey then goes from the 4:30 position ($\theta = -45°$)

to the 1:30 position ($\theta = +45°$). The starting velocity is v in the 1:30 direction; the ending velocity is v in the 10:30 direction (Figure 6.5a).

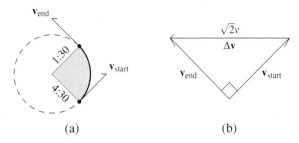

(a)   (b)

Figure 6.5 Calculating $\Delta\mathbf{v}$ for the quarter-revolution journey. (a) The runner's starting and ending velocities. (b) $\Delta\mathbf{v}$ as their vector difference (tip-to-tip).

Their (vector) difference $\Delta\mathbf{v}$ is calculated by drawing the two vectors with a common tail; the difference is the vector from $\mathbf{v}_{\text{start}}$'s tip to $\mathbf{v}_{\text{end}}$'s tip (Figure 6.5b). Thus, $\Delta\mathbf{v}$ is the hypotenuse of the right triangle with $\mathbf{v}_{\text{start}}$ and $\mathbf{v}_{\text{end}}$ as its legs. Each leg has length v, so $\Delta\mathbf{v}$ has magnitude $\sqrt{2}v$. With Δt from (6.47),

$$\mathbf{a}_{\text{avg}} \equiv \frac{\Delta\mathbf{v}}{\Delta t} = \frac{\sqrt{2}v \text{ to the left}}{\pi R/2v} = \frac{2\sqrt{2}}{\pi}\frac{v^2}{R} \text{ to the left.} \qquad (6.48)$$

Like the first average acceleration (6.46), for which Δt was one-half of a period, this average acceleration points to the left, contains a factor v^2/R, and has a dimensionless constant. In the first situation, however, the dimensionless constant was $2/\pi \approx 0.64$. In this second situation, it's $2\sqrt{2}/\pi \approx 0.90$.

In the third situation, the runner journeys now only one-sixth of a revolution (Figure 6.6a), from 4 o'clock ($\theta = -30°$) to 2 o'clock ($\theta = +30°$) with 3 o'clock still the midpoint. Then $\mathbf{v}_{\text{start}}$ points in the 1 o'clock direction, and $\mathbf{v}_{\text{end}}$ points in the 11 o'clock direction.

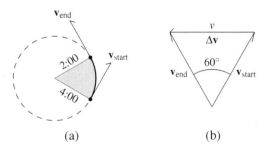

(a)   (b)

Figure 6.6 Calculating $\Delta\mathbf{v}$ for the one-sixth-revolution journey. (a) The runner's starting and ending velocities. (b) $\Delta\mathbf{v}$ as their vector difference. The starting and ending velocities and their difference form an equilateral triangle.

These orientations differ by $60°$ (one-sixth of a revolution). Because $\mathbf{v}_{start}$ and $\mathbf{v}_{end}$ have the same length (the constant speed v), the three vectors $\mathbf{v}_{start}$, $\mathbf{v}_{end}$, and $\Delta\mathbf{v}$ form an equilateral triangle (Figure 6.6b). Thus,

$$\Delta\mathbf{v} = v \text{ to the left.} \tag{6.49}$$

With this $\Delta\mathbf{v}$ and with Δt equal to one-sixth of a period (to $\pi R/3v$),

$$\mathbf{a}_{avg} = \frac{v \text{ to the left}}{\pi R/3v} = \frac{3}{\pi}\frac{v^2}{R} \text{ to the left.} \tag{6.50}$$

The structure of (6.46), for one-half of a period, and of (6.48), for one-third of a period, recurs. The average acceleration points to the left, contains the symbolic factor v^2/R, and has a dimensionless constant. For one-half of a period, the constant is $2/\pi \approx 0.64$. For one-third of a period, it's $2\sqrt{2}/\pi \approx 0.90$. For one-sixth of a period, it's $3/\pi \approx 0.95$.

▷ *Can you guess what happens to the average acceleration as Δt goes to zero?*

As Δt has shrunk, the constant has become closer to 1. So, a reasonable guess is that, as Δt goes to zero, the constant becomes exactly 1. Then the acceleration would become simply v^2/R to the left.

The physicist John Wheeler's first moral principle is, "Never make a calculation until you know the answer" [22, p. 20]. So, only now, with the preceding guess in hand, should we calculate the instantaneous acceleration directly.

From (6.41), the instantaneous acceleration is the average acceleration in the limit $\Delta t \to 0$. Thus, imagine a very short time interval Δt, measured as a fraction f of a full period. In that time, the runner journeys along an arc of angle $\Delta\theta = 360f°$ or $2\pi f$ radians (Figure 6.7a). With the f notation, the three average-acceleration situations were $f = 1/2$, $f = 1/4$, and $f = 1/6$; our goal now is that f go to zero.

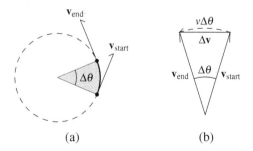

(a)   (b)

Figure 6.7 Calculating $\Delta\mathbf{v}$ for an infinitesimally short journey spanning an angle $\Delta\theta$. (a) The runner's starting and ending velocities. (b) $\Delta\mathbf{v}$ has approximate magnitude $v\Delta\theta$: a hair shorter than the dashed arc, whose length is exactly $v\Delta\theta$.

The easier part of the computation is the duration Δt. Using the full period T calculated in (6.44),

$$\Delta t = fT = \frac{2\pi f R}{v}. \tag{6.51}$$

The harder part is the velocity difference $\Delta \mathbf{v}$ (Figure 6.7b). The starting and ending velocity vectors have length v (the constant speed) and are almost vertical. Their difference, which points to the left, depends on how close to vertical they are, which in turn depends on the angle between them. Because the velocity vectors are tangent to the circle, they rotate by the same angle as subtended by the runner's arc – which is $\Delta\theta$ or $2\pi f$ radians.

When $\Delta\theta$ is tiny, the horizontal side of the triangle, which represents the chord $\Delta\mathbf{v}$, has almost the same length as the arc of the circle that surrounds $\Delta\mathbf{v}$. This arc's length is exactly $v\Delta\theta$. With $\Delta\theta = 2\pi f$, the length of the chord $\Delta\mathbf{v}$ is approximately $2\pi f v$. Thus,

$$\Delta\mathbf{v} \approx 2\pi f v \text{ to the left.} \tag{6.52}$$

To check this approximation, we can test it on $f = 1/6$ (the third average-acceleration calculation). The approximation predicts

$$\Delta\mathbf{v} \approx \frac{2\pi}{6} v \text{ to the left.} \tag{6.53}$$

When $f = 1/6$, the exact $\Delta\mathbf{v}$, in (6.49), is v to the left. Because $2\pi \approx 6$, the approximate $\Delta\mathbf{v}$ in (6.53) and the exact $\Delta\mathbf{v}$ agree enough that I trust the approximation (6.52).

With $\Delta\mathbf{v}$ from (6.52) and with $\Delta t = 2\pi f R / v$ from (6.51),

$$\mathbf{a}_{\text{avg}} \equiv \frac{\Delta\mathbf{v}}{\Delta t} \approx \frac{2\pi f v \text{ to the left}}{2\pi f R / v}. \tag{6.54}$$

Many factors cancel: the 2, the π, and the f. What remains is v^2/R. Thus,

$$\mathbf{a}_{\text{avg}} \approx \frac{v^2}{R} \text{ to the left.} \tag{6.55}$$

As f goes to zero (which gives the limit $\Delta t \to 0$), two effects happen. First, the approximation embodied in (6.52), that the $\Delta\mathbf{v}$ side is the arc of the circle, becomes exact. Second, the average acceleration $\mathbf{a}_{\text{avg}}$ becomes the instantaneous acceleration $\mathbf{a}$. Thus,

$$\mathbf{a} = \frac{v^2}{R} \text{ to the left.} \tag{6.56}$$

The direction, to the left, applies to runner at the 3-o'clock position ($\theta = 0$). At a general position along the circle, "to the left" translates into "toward the center." To summarize this great result (and answer the triangle question):

> **Uniform Circular Motion.** For a particle moving in a circle of radius R at constant speed v,
>
> $$\mathbf{a} = \frac{v^2}{R} \text{ toward the center.} \qquad (6.57)$$

Fortunately after all that work, this formula makes physical sense. The higher the speed (for a fixed radius), the faster the particle rounds the circle and the faster the velocity changes orientation – and, therefore, the larger the acceleration magnitude. Thus, v should, and does, appear in the numerator. In contrast, the larger the radius (for a fixed speed), the slower the velocity changes orientation. Therefore, R should, and does, appear in the denominator.

The acceleration (6.57) never points in the same direction as the velocity! Thus, in circular motion and in general two-dimensional motion, distinguishing acceleration from velocity becomes essential. For if you incorrectly connect force to velocity rather than to acceleration or correctly connect it to acceleration but incorrectly assume that the acceleration points in the same direction as the velocity, you will be tempted to invent bogus forces in the direction of motion. When that happens, Newton's laws disappear out the window – along with your chances of understanding the system.

6.4 Constant-Speed Motion around an Ellipse

In the analysis of circular motion (Section 6.3), two parameters remained constant throughout the motion: the particle's speed and its distance from the center. The next analysis removes the constant-distance constraint: We study motion around an elliptical path (but at constant speed).

Thus, imagine a racing or train car moving at constant speed around a large elliptical track (Figure 6.8). The track's long radius r_{max} (the semimajor axis) is 2 kilometers, and its short radius r_{min} (the semiminor axis) is 1 kilometer.

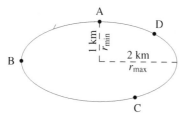

Figure 6.8 An elliptical track around which a car moves at constant speed.

> *Can you draw the acceleration vector at each of the four indicated positions, getting the direction and relative magnitude reasonably correct?*

At point A, the car is r_{min} from the center of the ellipse. But beware: If we blindly calculate the acceleration using (6.57), we will conclude that the acceleration has magnitude v^2/r_{min} and points toward the center (which, at point A, is downward). That conclusion is only half right.

To see what part is wrong, imagine an easy extreme case: that the elliptical path becomes even more elliptical. For example, imagine stretching the ellipse horizontally so that r_{max} grows to 1000 kilometers (while r_{min} remains fixed at 1 kilometer). Then the elliptical path around point A straightens out almost to a straight line. And motion in a straight line at constant speed means zero acceleration. So, stretching the ellipse should make the acceleration magnitude go to zero. Yet, as r_{max} grows, the predicted acceleration magnitude v^2/r_{min} doesn't even change. So, v^2/r_{min} cannot be the acceleration magnitude.

To make an analogous extreme case, squash the ellipse vertically by holding r_{max} fixed while shrinking r_{min} to zero. As the ellipse squashes, the path around point A again straightens out, so the acceleration magnitude should again go to zero. But the prediction v^2/r_{min} goes to infinity! With two strikes against it, it cannot be the correct acceleration magnitude.

But it's almost correct. We have to replace R, the particle's distance from the center, with the path's radius of curvature $r_{curvature}$: the radius of the circle that best kisses the path around point A (Figure 6.9a). For a circle (Section 6.3), the radius of the kissing circle always equals the radius of the circular path itself; thus, we didn't need to distinguish the two concepts and could use R to stand for either. For an ellipse, however, the radius of curvature varies along the path.

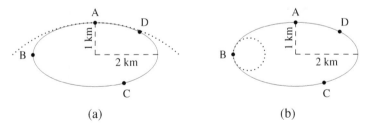

(a)   (b)

Figure 6.9 Measuring radius of curvature. The circle that curves just like a path does at a particular point is the kissing circle. Its radius is the path's radius of curvature there. (a) The kissing circle (dotted) at point A. (Only a portion of the circle is drawn because it's huge.) Its radius is 4 kilometers. (b) The kissing circle at point B. This circle, with radius 0.5 kilometers, is tiny compared to the kissing circle at point A, reflecting the straighter path at point A compared to at point B.

To find the acceleration, we approximate the path at point A using a kissing circle (Figure 6.9a) and then follow the circular-motion analysis of Section 6.3. The result, from (6.57), is that the acceleration has magnitude

$$a = \frac{v^2}{r_{\text{curvature}}} \tag{6.58}$$

and points toward the center of the approximating circle – that is, it points inward, perpendicularly to the particle's velocity (its direction of motion).

In this problem, the numerator v^2 never changes (the car moves at constant speed), but $r_{\text{curvature}}$ varies. The path is the straightest at point A, making the approximating circle's radius $r_{\text{curvature}}$ the largest there. So, a is the smallest at point A.

At point B, where the path is the most curved, the approximating circle's radius is the smallest (Figure 6.9b). So, a will be the largest there. And at points C and D, the accelerations will have intermediate magnitudes. These results (Figure 6.10) answer the triangle question. On this diagram and on subsequent freebody diagrams, the acceleration vectors are barbed only on a single side (like velocity vectors would be) – but with two harpoons (a reminder that acceleration, in calculus terminology, is the second derivative of position).

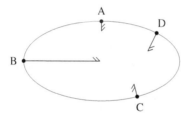

Figure 6.10 The acceleration at points A, B, C, and D. Because the car's speed is constant, the acceleration points directly inward. Its magnitude, inversely proportional to $r_{\text{curvature}}$, is the greatest at point B, where the path has the smallest radius of curvature (is the most curved), and is the smallest at point A.

Around the whole elliptical track, the acceleration is directly inward – that is, it's perpendicular to the track. This apparent coincidence arises because the car's speed is constant. In the next example (Section 6.5), we investigate how the acceleration is affected by a changing speed.

6.5 Varying-Speed Motion around a Circle

Before studying the full complexity of motion – noncircular motion with a varying speed (Section 6.6) – we'll revisit motion in a circle but allow the speed to vary. The example is a pendulum bob (Figure 6.11) shown at five points in its swing from one extreme (point A) to the other (point E). The bob's velocity, at least, is familiar from everyday experience. In direction, it always points along

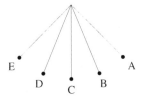

Figure 6.11 A pendulum bob swinging forever between one extreme (point A) and the other (point E). The bob moves along a circular path but with a varying speed.

the circle. In magnitude, it's zero at either extreme and is largest at the middle, at point C (the equilibrium position). But, as you have read many times in this book, velocity ain't acceleration! And the acceleration, being the rate of change of velocity, is usually harder to compute than is the velocity.

Can you draw, roughly, the acceleration vector at each point?

This question, from research by the physics educators Frederick Reif and Sue Allen [18, 19], gave physics students and teachers terrible headaches; hardly any students and few teachers solved it correctly. But, if you keep your wits about you – that is, reason from first principles – you can solve it.

Determining the acceleration is the easiest at point A, as long as you don't confuse zero velocity with zero acceleration. Instead, make a thought experiment to compute the average acceleration over a short interval: Imagine the bob swinging up toward point A. Once it reaches point A, start the clock. Slightly later, when the bob is moving along the circle away from point A, stop the clock.

The initial velocity vector, v_{start}, is zero: The bob is at rest at either extreme. (The bob being at rest makes a point an extreme.) The final velocity vector, v_{end}, points away from point A almost exactly along the circle. Thus, their difference Δv also points along the circle away from point A, as does the average acceleration, $\Delta v / \Delta t$. Similarly, at point E, the other extreme, the average acceleration points along the circle and away from point E. At either extreme, the bob falls down the circle as if it were sliding down an inclined plane.

The next-easiest calculation is at point C, the equilibrium position. There are two possibilities for the motion, which turn out (not by chance) to give the same acceleration. Namely, the bob could move to the left or to the right. The following analysis is for a left-moving bob. In Problem 6.4, you confirm that a right-moving bob has the same acceleration.

Thus, imagine a left-moving bob just before and just after point C, at locations symmetric around point C. Just before point C, the velocity, tangent to the circle, points mostly to the left and slightly downward (Figure 6.12a). Just after point C, the velocity points mostly to the left but now slightly upward.

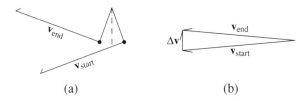

(a)   (b)

Figure 6.12 Calculating $\Delta \mathbf{v}$ and $\mathbf{a}_{avg}$ around point C. (a) The starting and ending velocities. By symmetry, these velocities have the same magnitude. (b) Their difference $\Delta \mathbf{v}$ is a tiny upward, or inward, vector, as is therefore $\mathbf{a}_{avg}$.

Because of symmetry, these before-and-after speeds are identical. Thus, while the bob is near C, the bob acts just like a particle moving around a circle at constant speed – motion whose acceleration you know (Section 6.3)! From (6.57), the acceleration points inward – which, at point C, means upward (Figure 6.12b) – and has magnitude $a = v^2/R$, where R is the radius of the circular path (here, the length of the pendulum string).

At point B, the acceleration has features of the acceleration at point A (that the speed changes) and at point C (that the direction of motion changes). To see how they combine, let's analyze a left-moving bob. (In Problem 6.5, you analyze a right-moving bob.) Just before point B, the velocity, always tangent to the circle, points downward and to the left. In comparison, the velocity just after point B points slightly more upward (Figure 6.13a). If this change of direction were the whole $\Delta \mathbf{v}$, then the acceleration would point inward (along the string) and have magnitude v^2/R (where v is the bob's speed at point B).

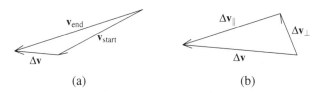

(a)   (b)

Figure 6.13 $\Delta \mathbf{v}$ around point B. (a) Calculating $\Delta \mathbf{v}$. As at point C, the ending velocity is rotated clockwise relative to the starting velocity. At point B, however, it also lengthens (as the bob "falls" downhill). (b) Splitting $\Delta \mathbf{v}$. The velocity difference $\Delta \mathbf{v}$ can be split into two portions. The parallel portion $\Delta \mathbf{v}_{\parallel}$ is due to the change in speed. The perpendicular portion $\Delta \mathbf{v}_{\perp}$ is due to the change in direction.

However, the story isn't complete. For the bob speeds up as it travels downhill along the circular path; in calculus terms, $dv/dt > 0$. Thus, this left-moving bob moves faster after point B than it does before point B. So, the velocity not only changes direction, it also changes in magnitude.

The change in speed produces $\Delta \mathbf{v}_{\parallel}$: the portion of $\Delta \mathbf{v}$ along the direction of motion (along $\mathbf{v}$). The change in direction produces $\Delta \mathbf{v}_{\perp}$: the portion of $\Delta \mathbf{v}$ perpendicular to the direction of motion (Figure 6.13b). The two portions, after dividing by the duration Δt, make the two corresponding portions of the acceleration vector, $\mathbf{a}_{\parallel}$ and $\mathbf{a}_{\perp}$. In symbols, $\mathbf{a} = \mathbf{a}_{\parallel} + \mathbf{a}_{\perp}$, where

$$\mathbf{a}_{\parallel} = \frac{dv}{dt} \text{ along } \mathbf{v}, \text{ and}$$

$$\mathbf{a}_{\perp} = \frac{v^2}{R} \text{ inward.}$$

(6.59)

At point B, $\mathbf{a}_{\parallel}$ points, just like the velocity, mostly left and slightly downward. Meanwhile, $\mathbf{a}_{\perp}$ points inward (along the string). Their sum – the acceleration $\mathbf{a}$ – ends up mostly left and slightly upward (Figure 6.14). Calculating the exact direction requires knowing the speed at point B, a calculation that I haven't explained (because it depends on energy conservation). For that reason, I asked you only for the rough direction of the acceleration vectors. But Figure 6.14 shows the exact directions and (relative) lengths.

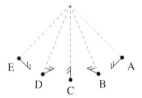

Figure 6.14 The acceleration vector at the given points. At points A and E, the bob is changing its speed but, because $v = 0$, not its direction, so the acceleration vector is parallel to the velocity. At point C, the bob is changing its direction but not its speed, so the acceleration vector is perpendicular to the velocity and thus inward. At points B and D, the acceleration vector is a mixture of these two cases.

Knowing the acceleration at point B ($\mathbf{a}_B$), we can find the acceleration at the remaining point, point D, by symmetry. Just as $\mathbf{a}_A$ and $\mathbf{a}_E$ are mirror images of one another (mirrored in the vertical plane through point C), $\mathbf{a}_B$ and $\mathbf{a}_D$ are also mirror images. The mirror negates only the horizontal portion of the acceleration and leaves the vertical portion alone. Thus, because $\mathbf{a}_B$ points mostly leftward and slightly upward, $\mathbf{a}_D$ points mostly rightward but still slightly upward.

This symmetry argument also shows that $\mathbf{a}_C$ points purely upward. Point C is its own mirror image. Thus, $\mathbf{a}_C$ is also its own mirror image, so it has no horizontal portion (whether leftward or rightward).

In summary and in answer to the triangle question: Each acceleration vector has a perpendicular portion that, when nonzero, always points inward; and each has a parallel portion that depends on whether the particle is speeding up (the parallel portion then points forward), is slowing down (the parallel portion then points backward), or is at a speed maximum or minimum (the parallel portion is then zero). The acceleration vector is the sum of these two portions.

6.6 Acceleration in General: A Summary

The result (6.59) for the parallel and perpendicular portions of the acceleration is so important that I now summarize the preceding analyses in order to emphasize their key points.

1. Velocity **v**, which measures the rate at which a body's position changes with time, is a vector. Therefore, it has magnitude (the speed) and direction.

2. Acceleration **a**, also a vector, measures the rate at which a body's velocity changes with time.

3. Velocity, like any vector, can change because its magnitude changes or because its direction changes.

4. Acceleration can be split into two portions corresponding to the two ways in which velocity can change (point 3).

5. Changing speed (the first way that velocity can change) produces $\mathbf{a}_\parallel$, the parallel portion of the acceleration vector. This portion lies in the same line as the body's velocity and has magnitude $|dv/dt|$, where dv/dt is the rate at which the body's speed changes. More precisely,

$$\mathbf{a}_\parallel = \frac{dv}{dt} \text{ along } \mathbf{v}. \qquad (6.60)$$

   Using $\mathbf{a}_\parallel$, you can translate into unambiguous mathematics our everyday language about "accelerating" and "decelerating" that would otherwise create confusion. An accelerating body has an increasing speed v. In mathematical language, $dv/dt > 0$, and $\mathbf{a}_\parallel$ points forward; it's parallel to the velocity **v**. In contrast, a decelerating body has $dv/dt < 0$, and $\mathbf{a}_\parallel$ points backward; it's antiparallel to the velocity. Equivalently, $a_\parallel$, the component of **a** along the velocity, is dv/dt. (If **v** = 0, then **v** has no defined direction. In that case, imagine **a** at an instant just before or after **v** becomes 0, as we did for the pendulum in Section 6.5 at the extreme points A and E.)

6. Changing direction (the second way that velocity can change), meaning that the body is following a curved path, produces $\mathbf{a}_\perp$, the perpendicular portion of the acceleration vector. This portion points perpendicularly to the path and inward (toward the center of the kissing circle). It has magnitude $v^2/r_{\text{curvature}}$, where $r_{\text{curvature}}$ is the radius of the kissing circle (the circle that matches the path at a particular point). Thus,

$$\mathbf{a}_\perp = \frac{v^2}{r_{\text{curvature}}} \text{ inward.} \qquad (6.61)$$

▶ *On a curved path, "inward" has a clear meaning, which makes it possible to orient $\mathbf{a}_\perp$. But when the path is straight, how do you know which direction is inward?*

A straight path has $r_{\text{curvature}} = \infty$, making $v^2/r_{\text{curvature}}$, the magnitude of the perpendicular acceleration, zero. Thus, it doesn't matter in which direction the perpendicular acceleration points – because it doesn't exist: $\mathbf{a}_\perp = 0$.

7. Even when a body is, in everyday language, not accelerating – neither speeding up nor slowing down – it can still, in physics language, have a nonzero acceleration if it is changing its direction of motion.

These points are summarized in Figure 6.15.

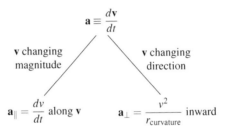

$$\mathbf{a} \equiv \frac{d\mathbf{v}}{dt}$$

$\mathbf{v}$ changing magnitude $\mathbf{v}$ changing direction

$$\mathbf{a}_\| = \frac{dv}{dt} \text{ along } \mathbf{v} \qquad \mathbf{a}_\perp = \frac{v^2}{r_{\text{curvature}}} \text{ inward}$$

Figure 6.15 Splitting acceleration into its parallel and perpendicular portions. $\mathbf{a}_\|$ arises from $\mathbf{v}$ changing magnitude; it points either in or opposite to the direction of the velocity. $\mathbf{a}_\perp$ arises from $\mathbf{v}$ changing direction; it points inward.

Now that acceleration's mysteries and subtleties are less so, we can restore acceleration to Newton's second law and use the law in its general form – the subject of Chapter 7.

6.7 Problems

6.1 In traveling from point A to point B (Figure 6.16), a particle moves at constant speed in the two straight segments: at 10 meters per second before reaching the semicircle and at 12 meters per second after leaving the semicircle (and speeds up steadily in the semicircle, as in Figure 6.2). For this journey:

a. Which will be larger, $|\mathbf{v}_{\text{avg}}|$ (the magnitude of the average velocity) or v_{avg} (the average speed)? Make your choice without calculating.

b. Check your expectation by calculating $\mathbf{v}_{\text{avg}}$, $|\mathbf{v}_{\text{avg}}|$, and v_{avg}.

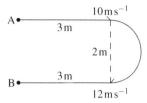

Figure 6.16 The particle of Problem 6.1 moving from point A to point B via the two straightaways and the semicircular track.

6.2 For the following situations, give (i) the acceleration **a** (magnitude and direction), (ii) its magnitude a, and (iii) its x component a_x. The positive x direction is to the right.

a. A train is moving at 100 meters per second to the left and is steadily gaining 1 meter per second of speed every second. (For example, 5 seconds later, its speed is 105 meters per second.)

b. A train is moving at 100 meters per second to the left and is steadily losing 2 meters per second of speed every second.

c. A car is moving at 50 meters per second to the right and is steadily gaining 3 meters per second of speed every second.

d. A car is moving at 50 meters per second to the right and is steadily losing 3 meters per second of speed every second.

e. A ball is moving upward at 20 meters per second and is steadily losing 10 meters per second of speed every second.

6.3 To high accuracy, the earth orbits the sun in a circle at constant speed, which we calculated (as the average speed) in (6.14). Find the earth's orbital acceleration (in magnitude and direction).

6.4 Redo the analysis of the acceleration at point C for a right-moving bob. Confirm that the acceleration still points upward.

6.5 Redo the analysis of the acceleration at point B for a right-moving bob. Confirm that $\mathbf{a}_\parallel$ still points mostly leftward and slightly downward. That is, now it should point *opposite* to the velocity. Check that this conclusion is consistent with $\mathbf{a}_\parallel$ as given in (6.59): For the right-moving bob at point B, what's the sign of its dv/dt?

6.6 A car salesman might claim, "This baby here can go from 0 to 60 in 4 seconds flat." Assuming that the baby is a car, that 60 represents 60 miles per hour, and that "flat" means "with no decimal point," translate that claim into a statement about the car's (average) acceleration. (In case you didn't grow up with miles: 60 miles per hour approximately equals 100 kilometers per hour.)

7

Newton's Second Law with Changing Motion

Chapter 5 introduced Newton's second law in the special case of statics: where the acceleration, and therefore the net force, was zero. Chapter 6 introduced acceleration alone, to describe changing motion. This chapter restores Newton's second law to its full glory connecting acceleration to its cause, net force.

7.1 Restoring Acceleration to Newton's Second Law

Newton's second law, in my preferred form (4.6), reminds of three important points. (1) Through the "net" subscript: The force on the left side is the net force, the sum of all forces acting on the body. (2) Through the boldface $\mathbf{F}$ and $\mathbf{a}$: The net force and acceleration are vectors. (3) Through the direction of the causal arrow ($\rightarrow$): Net force causes acceleration. In the rest of this section, you practice the full second law on motion in one dimension – including on the subtle process by which a passive force adjusts itself (Section 7.1.6).

In Section 7.2, you learn how to extend the second law to a composite body: whether a multipart body (a galaxy consisting of billions of stars) or an extended, rigid body (a stick). In Section 7.3, we turn to two-dimensional motion. Finally, in Section 7.4, we combine the preceding ideas to understand weight: how it's different from the gravitational force and how this distinction prevents many common misconceptions. The section closes with a solution to the bumblebee problem introduced in Section 2.3.

7.1.1 Falling Stone

Our simplest example of the full second law is a body acted on by a single, constant force: a stone starting from rest and falling without air resistance. This stone's only interaction is its gravitational interaction with the earth. Thus, the only force acting on it is the gravitational force, mg downward – which is then also the net force, $\mathbf{F}_{net}$. The stone is in free gravitational motion (free fall).

The following Newtonian analysis is of type C (calculating). Knowing the forces on the stone completely, we use this knowledge and the second law to calculate the stone's motion.

The first step is finding the stone's acceleration. From the second law (4.6),

$$\mathbf{a} = \frac{\overbrace{mg \text{ downward}}^{\mathbf{F}_{net}}}{m} = g \text{ downward}. \tag{7.1}$$

> *Given that the stone starts from rest, what's its speed v after a time t?*

Acceleration is the rate of change of velocity. Thus, the stone's velocity starts at zero and increases at a rate of g downward. Thus, $\mathbf{v} = gt$ downward, and $v = gt$.

> *How far does the stone fall in the time t?*

Based on the definition of average speed (6.13), the distance fallen is t multiplied by the stone's average speed. The stone's speed increases steadily from 0 to gt. Thus, its average speed is $gt/2$. So,

$$\text{distance fallen} = \underbrace{\text{time}}_{t} \times \underbrace{\text{average speed}}_{gt/2} = \frac{1}{2}gt^2. \tag{7.2}$$

This calculation applies more widely than just to this stone. Any freely falling body has the same freebody diagram with only a gravitational force (mg downward), has the same acceleration (g downward), and – if released from rest – falls the same distance $gt^2/2$ in the time t.

Surprisingly, the acceleration, velocity, and distance fallen are all independent of mass: The gravitational force is proportional to mass, but the $1/m$ factor in the second law cancels it out from the acceleration and, therefore, from the velocity and distance fallen.

However, a further surprise lurks behind that argument. For the cancellation of m is itself puzzling. The m in Newton's second law (in the $1/m$ factor) is a body's inertial mass: how strongly it resists changes in acceleration. However, the other m – the m in mg – is the body's gravitational mass: how strongly it's affected by gravity (here, from the Earth). Newtonian mechanics provides no reason for these two masses to be equal. But Einstein wondered whether there might be a deep reason. He found one, and it became the basis of general relativity, Einstein's theory of gravity. You will meet it briefly in Section 8.3.2. But here, back in the Newtonian world, we just accept the equality of these two kinds of mass as an amazing cosmic coincidence.

When a piece of toast falls from a dining table, how long do you have to catch it before it hits the ground (probably butter-side down)?

A standard table, judging from one nearby, has height $h \approx 0.6$ meters (2 feet). Inverting the distance fallen (7.2) to solve for t,

$$t = \sqrt{\frac{2h}{g}} \approx \sqrt{\frac{2 \times 0.6\,\mathrm{m}}{10\,\mathrm{m\,s^{-2}}}} \approx 0.35\,\mathrm{s}. \tag{7.3}$$

Human reaction times are 0.2 to 0.3 seconds, so our estimated time interval is, with a bit of luck, just long enough for us to react and catch the toast.

7.1.2 Ice Block Sliding on a Frozen Lake

Providing our second example of the full second law, a block of ice sits on a frozen lake (and on an airless world, allowing us to neglect air resistance). You kick the block, and it slides away from you (Figure 7.1), slows down, and eventually stops. To practice tracking minus signs and navigating conceptual sandbars, you kick it in the negative x direction (to the left).

Figure 7.1 A block of ice moving to the left after getting kicked.

How do Newton's laws describe the process of slowing down?

The first step in applying Newton's laws is making a freebody diagram. So that all the relevant information ends up in one place, the diagram should include the block's velocity and acceleration. The velocity is easy: As indicated in Figure 7.1 (which is not a freebody diagram), it points to the left. The acceleration is trickier. We might mistakenly conclude that, because the block is decelerating, its acceleration is negative and therefore points in the negative x direction.

What's wrong with the preceding argument?

The argument confuses velocity the vector with speed the scalar. It also confuses the everyday meaning of acceleration ("That motorcycle has some acceleration!") with the physics meaning of acceleration (as the acceleration vector). To clarify these distinctions, we'll compute the block's average acceleration, from kick to stop. From (6.22),

$$\mathbf{a}_{\text{avg}} \equiv \frac{\mathbf{v}_{\text{end}} - \mathbf{v}_{\text{start}}}{\Delta t}, \tag{7.4}$$

where $\mathbf{v}_{\text{start}}$ is the block's velocity just after your kick, $\mathbf{v}_{\text{end}}$ is its velocity once it stops, and Δt is the time span from your kick until the block stops.

At the start, just after your kick, the velocity is v_0 to the left, where v_0 is the block's initial speed. At the end, the velocity is zero. Thus,

$$\mathbf{a}_{\text{avg}} = \frac{0 - (v_0 \text{ to the left})}{\Delta t} = -\frac{v_0}{\Delta t} \text{ to the left.} \tag{7.5}$$

Because v_0 (a magnitude) and Δt are positive, the average acceleration is a negative amount, $-v_0/\Delta t$, to the left; equivalently, it's a positive amount, $+v_0/\Delta t$, to the right. Thus, the average acceleration points to the right, in the positive direction – even though the block is, in everyday language, decelerating. For reasons that become clear after we determine the forces, the block's acceleration remains constant while it slows down, so its average acceleration and (instantaneous) acceleration are identical:

$$\mathbf{a} = \mathbf{a}_{\text{avg}} = \frac{v_0}{\Delta t} \text{ to the right.} \tag{7.6}$$

This conclusion is consistent with two key points about acceleration discussed in Section 6.6. First (point 5), decelerating means that dv/dt, the rate of change of *speed*, is negative; and, therefore, that $\mathbf{a}_{\parallel}$, the parallel portion of the acceleration vector, points opposite to the velocity. Second (point 6), in one-dimensional or straight-line motion, the acceleration has no perpendicular portion $\mathbf{a}_{\perp}$. With $\mathbf{a}_{\perp} = 0$, the entire acceleration is $\mathbf{a}_{\parallel}$, which points opposite to the velocity. The velocity points in the negative x direction, so $\mathbf{a}$ points in the positive x direction – as we found in (7.6).

Knowing the correct acceleration, you can safely use Newton's second law to explain it. The explanation – the cause – is always net force: the sum of all forces acting on the block. In contrast, if you try to explain an incorrect acceleration, the analysis easily degenerates into a tangle of incorrect nonexistent forces.

Before finding the forces, I check with my intuitive model of the physical world and ask it how it describes the block's slowing down. This model is a mixture of two parts (Section 5.4.1). The ancient part reasons, "The block is moving to the left, so the (net) force on it must point left. Because the normal force and the gravitational force are vertical, they cannot help here. The left-pointing force must be another force, a force of motion. Because the block is slowing down, this force of motion must be decreasing in magnitude." The ancient part deeply but incorrectly believes that $\mathbf{F}_{\text{net}} \rightarrow m\mathbf{v}$.

To redirect it gently, I follow the advice of the physics educator Bob Kibble [11] and give it an alternative and more correct conceptual bone on which to chew. My preferred bone, for reasons that I explain below, is quantity of motion

– a physical quantity that depends on the body's velocity and its mass. Formally, this quantity is called momentum. Touched upon in Section 8.3.1, it's $m\mathbf{v}$, the product of mass and velocity.

The dependence on velocity acknowledges the ancient part's fascination with velocity. And this fascination is justified! For when we try to alter the world (engineering), acceleration is mostly a means to the end of changing a body's velocity until the velocity is what we want (which itself might be a means to the end of changing a body's position). In the other direction, when we try to predict how the world will behave (science) – for example, whether a body is going to hit us and, if so, whether it's dangerous – the most directly relevant information is again the body's velocity or speed but rarely its acceleration.

The dependence on mass acknowledges the ancient part's correct sense that more massive objects need more force to alter their behavior and often therefore present more danger.

With the new concept and name quantity of motion, I can redirect my intuitive part's attention away from the incorrect direct connection between motion and force and toward the correct indirect connection:

$$\text{net force} \quad \underbrace{\text{causes } (\Rightarrow)}_{\mathbf{F}_{\text{net}}} \text{changes} \quad \underbrace{\text{quantity of motion}}_{m\mathbf{v}}. \qquad (7.7)$$

The redirected ancient part can now correctly describe the block slowing down: "The block is moving to the left, so its quantity of motion points to the left. The block is slowing down, so the quantity is decreasing in magnitude." This statement not only won't confuse me as we formally analyze the forces, it can even help structure the analysis. When I hear "motion" in "quantity of motion" and how it points ever less to the left – in other words, how the body is acquiring a quantity of motion pointing to the right – I remember to expect a net force responsible for this change by pointing to the right (as long as I remember that quantity of motion is a vector). And I might well remember this conclusion despite the ice block's seemingly conflicting motion to the left.

(By using this new quantity, quantity of motion, and the idea that force changes it, the intuitive part actually leapfrogs beyond the formal idea that force changes velocity or causes acceleration. As long as a body's mass remains constant, the two descriptions of force are equivalent. However, when a body's mass changes – think of a rocket burning through its fuel – they are not equivalent. As you learn in Section 8.3.1, the quantity-of-motion description in (7.7) is then the correct one and is used in advanced physics. By redirecting your intuition toward quantity of motion, you prepare yourself lightly for advanced physics.)

But let's return to the sliding block and to the formal analysis of the forces acting on it. To find the forces and thereby complete the freebody diagram, return to the first principle of Newtonian physics – interaction – and to the resulting freebody-diagram recipe (Section 2.1). For the short-range interactions (step 1 of the recipe): The block touches only one body, the ice sheet (the frozen lake), so it participates in only one short-range interaction. This interaction contributes a contact force from the ice sheet (Figure 7.2a). Meanwhile, the block's only long-range interaction (step 2) is its gravitational interaction with the earth. This interaction contributes a gravitational force mg downward.

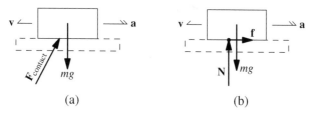

(a)   (b)

Figure 7.2 Freebody diagrams of the ice block (including its velocity and acceleration). (a) The diagram made from first principles. The block participates in only two interactions, so it experiences only two forces: gravity and the contact force from the icy lake. (The icy lake is shown in ghostly outline.) (b) The same diagram after splitting the contact force into its perpendicular portion (the normal force **N**) and its parallel portion (dynamic friction **f**).

The contact force, $\mathbf{F}_{contact}$, can be broken into two portions: the normal portion **N** perpendicular to the ice sheet and the parallel portion **f** along the ice sheet (Figure 7.2b). Because the block's acceleration has no vertical portion, the net force cannot have a vertical portion. Thus, the only vertical forces, **N** and the gravitational force, must balance. As a result, $N = mg$. Here, the analysis became partly type I (inferring): Knowledge of **a** (that it's purely horizontal) was used to infer the normal force.

The parallel portion **f** is a frictional force. Because the block moves relative to the ice sheet, the friction is dynamic friction (Section 1.3.4), which opposes the relative motion of the block and ice sheet. Thus, on the block, which moves to the left relative to the fixed ice, **f** points to the right. Furthermore, from (1.26),

$$f = \mu N = \mu mg, \tag{7.8}$$

where μ is the coefficient of dynamic friction. (Because of this relation, in which f depends on N, we compute **N** before **f**.) For an ice block sliding on ice, μ is tiny because the contact lubricates itself with a several-molecule-thick layer of meltwater (from ice melted by frictional heating).

In summary, and in answer to the original triangle question (p. 118), the Newtonian description of the slowing-down process has the following steps. (1) The block participates in two interactions (one short range and one long range); therefore, it's acted on by two forces: the short-range, contact force and the gravitational force. (2) The contact force can be broken into a perpendicular portion, the normal force $\mathbf{N}$; and a parallel portion, the dynamic friction $\mathbf{f}$. (3) Because the block's acceleration has no vertical portion, the net force cannot have any either. Thus, $\mathbf{N}$ balances the gravitational force, and $N = mg$. (4) The net force therefore equals $\mathbf{f}$, the only unbalanced force. (5) The dynamic friction $\mathbf{f}$ has magnitude μmg and points opposite to the block's velocity. (6) Thus, the block's acceleration points to the right, opposite to its velocity. (7) In other words, the block slows down (decelerates), and it eventually stops.

This description nowhere contains a force in the direction of motion. Thus, there cannot be and there isn't any "force of motion."

To demonstrate how forces, through the magic of the second law, determine a body's motion, and also to illustrate the type-C (calculate) part of the analysis, let's next make the description of the motion quantitative. We'll determine the following quantities about the block's motion: the block's acceleration, its stopping time Δt, its average speed while moving, and its stopping distance.

▶ *What's the block's acceleration?*

From steps 4 and 5 in the preceding summary,

$$\mathbf{F}_{net} = \mathbf{f} = \mu mg \text{ to the right.} \qquad (7.9)$$

(The usual mathematical notation, an equals sign, hides an important physical distinction: that $\mathbf{f}$, a portion of a physical force, is conceptually distinct from the net force. The net force need not equal any physical force at all – for example, it might be zero.) From the second law and the net force (7.9),

$$\mathbf{a} = \frac{\mathbf{F}_{net}}{m} = \mu g \text{ to the right.} \qquad (7.10)$$

Because μ and g are constant, so is the acceleration – which is why we could equate $\mathbf{a}_{avg}$ and $\mathbf{a}$ in (7.6). Interestingly, the acceleration, unlike the net force, does not depend on m: Although a more massive block experiences a larger dynamic-friction force, its larger mass makes it correspondingly harder to accelerate or decelerate. This independence of motion from mass occurs whenever the forces are, or are proportional to, gravitational forces; the simplest example is the stone of Section 7.1.1 in free gravitational motion (free fall). In such situations, all forces and the net force are proportional to m, a factor that then divides out of the acceleration because of the factor of $1/m$ in the second law.

▶ *What's the block's stopping time?*

Like a stone launched upward (the subject of Problem 7.3), the block's speed steadily decreases. The rate of decrease is $a = \mu g$. The block's initial speed is v_0. Thus, its stopping time Δt (the time for the speed to reach zero) is given by

$$\Delta t = \frac{\text{initial speed}}{\text{rate of decrease of speed}} = \frac{v_0}{\mu g}. \tag{7.11}$$

▷ *What's the block's average speed?*

Because **a** is constant, the block's speed decreases steadily with time (namely, at a constant rate), from v_0 to zero. Thus, its average speed is the halfway speed:

$$v_{\text{avg}} = \frac{1}{2}v_0. \tag{7.12}$$

(This argument, using a constant **a** to deduce a steady change in speed, works only for one-dimensional motion and, even then, only if the body doesn't reverse direction – see Problem 7.5. The complete condition, necessary and sufficient, for a steady change in speed is that $a_\parallel$, the component of $\mathbf{a}_\parallel$ along the body's velocity, be constant: As summarized in Section 6.6 (point 5), $a_\parallel$ is just dv/dt, the rate of change of speed.)

▷ *What's the block's stopping distance?*

The stopping distance d is the block's average speed times the stopping time:

$$d = v_{\text{avg}}\Delta t. \tag{7.13}$$

From (7.12), $v_{\text{avg}} = v_0/2$. Using the stopping time Δt from (7.11),

$$d = \frac{v_0}{2} \times \frac{v_0}{\mu g} = \frac{1}{2}\frac{v_0^2}{\mu g}. \tag{7.14}$$

▷ *What are rough but reasonable estimates for the preceding quantities?*

Let's say that you kick the ice block with speed $v_0 = 10$ meters per second – the speed of a fast sprint. For smooth ice sliding on smooth ice, the coefficient of dynamic friction μ is approximately 0.02. Then, from (7.11), the block's stopping time is approximately 50 seconds:

$$\Delta t = \frac{v_0}{\mu g} \approx \frac{10\,\text{m s}^{-1}}{0.02 \times 10\,\text{m s}^{-2}} = 50\,\text{s}. \tag{7.15}$$

From (7.12), the average speed is 5 meters per second:

$$v_{\text{avg}} = \frac{1}{2}v_0 = 5\,\text{m s}^{-1}. \tag{7.16}$$

From (7.14), the stopping distance is 250 meters:

$$d = \underbrace{5\,\text{m s}^{-1}}_{v_{\text{avg}}} \times \underbrace{50\,\text{s}}_{\Delta t} = 250\,\text{m}. \tag{7.17}$$

The block stops only after one-fourth of a kilometer!

Because the block's mass does not affect its acceleration, the mass also does not affect the stopping distance or time. The only parameters that affect the stopping distance and time are the gravitational strength g, which is the same for all objects on earth, the dynamic-friction coefficient μ (which depends on the materials involved), and the initial speed v_0. Thus, a car skidding on ice could have a comparably large stopping distance. Although tire rubber sliding on ice might have a higher μ than ice sliding on ice, which, per (7.14), would reduce the stopping distance compared to the ice block's, the car probably has a higher initial speed v_0, which would increase the stopping distance. Wise drivers in cold climates therefore fear black ice – ice so thin that it takes on the color of, and blends into, the road. You notice it only once you start skidding on it – too late to reduce v_0, which is better done when μ is higher (on the ice-free road).

7.1.3 Accelerating Bicycle

The next example of one-dimensional motion is an accelerating bicycle. In the form of its equations, it's similar to the kicked, decelerating ice block (Section 7.1.2). However, it contains an active agent, the cyclist, which adds a conceptual trap for the nonparanoid (most of us). To practice avoiding the traps, imagine yourself bicycling on level ground, as in Section 5.4, but accelerating. You and the bicycle – from now on considered as one composite body called the bicycle – move ever faster to the right, in the positive x direction. For simplicity, ignore air resistance, a reasonable simplification when you start from rest – say, after waiting at a stop sign (which few cyclists do, alas).

Our goal is a Newtonian description of the situation, one that connects the forces on the bicycle to the bicycle's acceleration. The description's first and essential step is a freebody diagram.

> *What's the freebody diagram of the bicycle?*

When making a freebody diagram, we first draw what we know about the body's velocity and acceleration – especially their directions. The velocity helps us orient any dynamic-friction or air-resistance forces. Here, the velocity points to the right. The acceleration, with the help of the second law, guides us toward a correct net force. Here, the acceleration also points to the right. Thus, the net force points to the right.

With that preparation, let's turn to the individual forces (whose sum is the net force). They arise from the bicycle's interactions. Like the sliding ice block (Section 7.1.2), the bicycle participates in one short-range (electromagnetic) interaction, with the ground, and one long-range interaction (gravitational), with the earth. Thus, the bicycle experiences two forces: a contact force $\mathbf{F}_{\text{contact}}$ and a gravitational force $\mathbf{F}_g$ (Figure 7.3a).

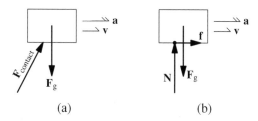

(a)   (b)

Figure 7.3 Freebody diagrams of the accelerating bicycle. (a) The diagram made from first principles. The bicycle participates in only two interactions, so it experiences only two forces: the gravitational force and the contact force from the ground. (b) The same diagram after splitting the contact force into its perpendicular portion (the normal force N) and its parallel portion (friction f).

$\mathbf{F}_{contact}$'s perpendicular portion is the normal force $\mathbf{N}$ (Figure 7.3b). Its parallel portion is the friction $\mathbf{f}$. With this partition of $\mathbf{F}_{contact}$, the second law (4.6) says

$$\underbrace{\frac{1}{m}\left(\mathbf{N} + \mathbf{f} + \mathbf{F}_g\right)}_{\mathbf{F}_{net}} \to \mathbf{a}. \tag{7.18}$$

As a vector equation,

$$\mathbf{N} + \mathbf{f} + \mathbf{F}_g = m\mathbf{a}. \tag{7.19}$$

Here, $\mathbf{N}$ and $\mathbf{F}_g$ are vertical, and $\mathbf{f}$ and $\mathbf{a}$ are horizontal. Based on that geometry, $\mathbf{N}$ and $\mathbf{F}_g$ must balance (add to zero) without any help from $\mathbf{f}$ or $\mathbf{a}$. Thus,

$$\mathbf{N} = -\mathbf{F}_g = mg \text{ upward.} \tag{7.20}$$

With that simplification, the vector equation (7.19) becomes

$$\mathbf{f} = m\mathbf{a} = ma \text{ to the right.} \tag{7.21}$$

Thus, the Newtonian description of this situation is that the bicycle is accelerated forward by the forward-pointing force of friction. But you might wonder:

Hasn't the force analysis omitted a force that the cyclist, as an active agent, exerts on the bicycle itself – and, therefore, on the combined system?

Invoking this force is the first conceptual trap of many in this seemingly simple problem. Its invocation is tempting because, without your effort, the bicycle would indeed just sit there (or fall over). So, you somehow cause the bicycle to accelerate. And acceleration, as I've now said *ad nauseam*, is caused by force. Thus, to restate the triangle question: Isn't the bicycle's acceleration caused by a force from you as the cyclist?

The flaw in the argument is that, whatever force you exert on the bicycle (now considering you and bicycle as two separate bodies), the bicycle exerts an equal and opposite force on you – by the third law. Thus, these two forces, as two sides of one interaction, cancel when we calculate the net force on the composite body of you and the bicycle – and only the net force determines the composite body's acceleration. (This important point gets further discussion in Section 7.2, on composite bodies.) Thus, the freebody diagram of the composite body should not have a force from you. Rather, as symbolized in (7.21), the composite body is accelerated by the frictional force from the road.

Meanwhile, a different forward-pointing force does appear in the interaction between you and the bicycle itself. However, it's the frictional force that the bicycle itself exerts on you: the same force that explains and causes your forward acceleration. From Newton's third law, therefore, you exert a *backward*-pointing frictional force on the bicycle. You try to accelerate the bicycle itself backward! But how then does the bicycle itself accelerate forward? The forward-pointing frictional force from the road more than overcomes the backward-pointing frictional force from you (Figure 7.4).

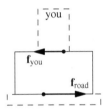

Figure 7.4 Frictional forces on the bicycle itself (these forces are the horizontal portions of their respective contact forces). You push the bicycle backward ($\mathbf{f}_{you}$), while the road pushes it forward ($\mathbf{f}_{road}$). The road's frictional force is stronger, so the bicycle accelerates forward.

Speaking of friction from the road, is it static or dynamic friction?

It depends on whether the tire is skidding. In the usual case, the tire does not skid. Then, at the point of contact, where the rubber touches the road (point C in Figure 5.11), the rubber does not move relative to the ground – just as for the nonaccelerating bicycle of Section 5.4. The rubber does move *relative to the hub*. It has just enough backward speed to balance and cancel the hub's forward motion relative to the ground. Thus, in the notation of (5.9), $\mathbf{v}_{C/hub} = 0$, so the friction is static friction. (Not realizing this point is the second conceptual trap.)

In the rarer case, you pedal so hard that you burn rubber. The bottom of the tire moves faster relative to the hub than the hub moves relative to the ground:

$$v_{C/hub} > v_{hub/ground}. \tag{7.22}$$

The two relative velocities fight, and, based on (7.22), the winner is $\mathbf{v}_{C/hub}$, which points backward. So, the bottom of the tire moves backward relative to the ground:

$$\underbrace{\mathbf{v}_{C/hub} + \mathbf{v}_{hub/ground}}_{\mathbf{v}_{C/ground}} \text{ points backward.} \qquad (7.23)$$

Because the tire and the ground move relative to each other, the friction is dynamic friction. Dynamic friction opposes the surfaces' relative motion, so the dynamic friction on the tire points forward. (Its counterpart force, which points backward, is the dynamic friction on the ground and easy to forget because the ground is so massive that this force produces almost no acceleration.)

When the friction is dynamic (burning rubber), its direction is decided by the relative motion of the ground and the tire. However, in the usual case of static friction, how does the frictional force determine its own direction?

The symmetry is broken by your pushing on the pedals in one direction rather than the other. Imagine your law-abiding self stopped at the stop sign, just before you start accelerating. To get going, you push down on the forward pedal, which rotates the chain, which rotates the (back) tire. As the wheel rotates, even ever so slightly, and before the hub and the whole bicycle move forward, the rubber at the contact point tries to move backward. Being pressed against the ground, which won't budge, the rubber cannot move. Instead, it compresses, as does the ground (though so slightly that you cannot easily see its compression). These new horizontal compressions (new deformations) combine with the tire's and the ground's existing vertical compressions and modify their existing spring or contact interaction (Figure 7.5).

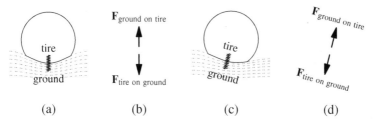

(a)   (b)   (c)   (d)

Figure 7.5 How pedaling changes the tire–ground spring interaction. (a) Before you pedal, the tire–ground compression is vertical. (b) This interaction consists of two vertical forces. (c) After you start pedaling (trying to move to the right), but before the hub has moved, the tire rotates clockwise, tilting the tire–ground spring interaction. (d) The result is two tilted spring (contact) forces. The horizontal portion of $\mathbf{F}_{ground \ on \ tire}$ is the static-friction force driving the bicycle forward.

Before the modification (that is, while you wait at the stop sign), the interaction contributes (1) a purely upward contact force on the tire and (2) its counterpart force, a purely downward contact force on the ground (Figure 7.5b). For conceptual simplicity, I pretend here that the bicycle has only one tire.

After the modification (that is, once you are pedaling), these two contact forces tilt (Figure 7.5d). The force on the tire tilts forward. The force on the ground, always equal and opposite, tilts backward. The new, forward-pointing portion of $\mathbf{F}_{\text{ground on tire}}$ *is* the static-friction force $\mathbf{f}$.

Equivalently, though slightly dangerously, you can think of the new horizontal compressions as creating a new, horizontal tire–ground interaction. (This picture is slightly dangerous because it implies that, for each tire, there are two tire–ground interactions, one vertical and one horizontal, even though there is really one tilted interaction, as shown in Figure 7.5c. But the picture is useful anyway because it corresponds to breaking up a contact force into perpendicular (or normal) and parallel (or frictional) portions.) This new, horizontal interaction contributes two new forces: (1) a backward force on the ground and (2) the counterpart force, a forward force on the tire. This forward force is the static-friction force $\mathbf{f}$. It represents the ground's compressed material fighting back as the material, like any self-respecting spring, tries to uncompress itself – by pushing forward the obnoxious tire, which so rudely tried to move backward.

As the rubber and ground compress further, the horizontal spring interaction and the resulting static-friction force strengthen, increasing the bicycle's acceleration. When the acceleration reaches what you want, you unconsciously adjust how hard you pedal in order to keep the compression at this value.

If the body is an accelerating car or train, which no one pedals, what process breaks the symmetry and decides the direction of static friction?

This question raises the third conceptual trap, that the active agent should be animate. In a car or train, the engine is the active agent, and the symmetry is broken by the direction that the engine tries to turn the wheels. Indirectly, this direction is decided by an animate being, the driver, who put the vehicle in forward or reverse gear. However, the engine does not directly cause the vehicle's acceleration. The engine is part of the vehicle, so it cannot be one side of an interaction *with* the vehicle. Thus, no "force from the engine" should appear on the freebody diagram for the vehicle. Rather, the vehicle's forward acceleration is caused by the ground through the vehicle–ground contact interaction. (This point is elaborated in Sections 7.2 and 7.3.2.)

7.1.4 Sliding Down a Frictionless Ramp

To generalize another no-acceleration analysis, return to that idyllic time when you stood peacefully on a hill (Section 5.2), when static friction balanced the parallel portion of gravity and held you in place. Now comes a prankster who lubricates the hill with a perfect oil, so perfect that all friction, static and dynamic, vanishes. You accelerate downhill.

What forces act on you, and what's your resulting acceleration magnitude?

The oil doesn't change the fundamental conclusion of the interaction analysis, that only two forces act on you: the gravitational force and the contact force. Because static friction has vanished, and static friction is the parallel portion of the contact force, the contact force must now be perpendicular to the hill – and stays that way even once you are sliding, because dynamic friction is also zero. This now-perpendicular force is called the normal force.

Furthermore, because you move only along the hill (and the hill is a straight line), your acceleration perpendicular to the hill ($\mathbf{a}_\perp$) must be zero. (To think more deeply about this argument, try Problem 7.8.) Thus, the net force cannot have a perpendicular portion. Therefore, the normal force must be balanced by the perpendicular portion of the only other force, the gravitational force (Figure 7.6). The net force is then just $\mathbf{F}_\parallel^g$, the parallel portion of the gravitational force. To state this argument more compactly: Of the four possible portions (two forces, each split into two portions), one portion ($\mathbf{f}$) vanishes and two portions ($\mathbf{N}$ and $\mathbf{F}_\perp^g$) balance, leaving only one portion ($\mathbf{F}_\parallel^g$) to act.

Figure 7.6 The forces on you after splitting the gravitational force $\mathbf{F}_g$ into its parallel portion $\mathbf{F}_\parallel^g$ and its perpendicular portion $\mathbf{F}_\perp^g$. Because you move in a straight line (here, downhill), your acceleration perpendicular to the ramp is zero. Thus, $\mathbf{F}_\perp^g$ must balance $\mathbf{N}$, and so $\mathbf{F}_{net} = \mathbf{F}_\parallel^g$.

This portion's magnitude is $mg \sin \theta$ (as it was when you stood on the unlubricated hill). Thus, in answer to the triangle question (about a),

$$\mathbf{a} = \underbrace{\frac{mg \sin \theta \text{ downhill}}{m}}_{\mathbf{F}_{net}} = \underbrace{g \sin \theta}_{a} \text{ downhill}. \qquad (7.24)$$

You accelerate downhill as if in free gravitational motion (free fall) but with the usual free-fall acceleration magnitude g multiplied by $\sin \theta$.

Multiplying g by a controlled factor less than 1 is a venerable theme in the development of physics. Even before George Atwood developed his machine (introduced in Section 5.6.3) to reduce, and thereby make measurable, the strong effects of gravity, Galileo had rolled balls down ramps. The ramp multiplied g by the sin θ factor that we just calculated in (7.24). And rolling itself (compared to frictionless sliding) multiplied g by a further factor of $5/7$ – a factor visible in Galileo's data and that can be calculated using torque (an idea touched upon in Section 8.2).

By reducing the acceleration, Galileo could study motion carefully and elucidate what acceleration means. The answer given in (6.41) – that **a** is the rate of change of velocity with respect to time – may seem evident to us today. However, it raises fundamental questions that Galileo had first to realize and then to answer. For example, what is velocity (which raises the Greek paradoxes of motion)? What does rate of change even mean? For example, is it measured with respect to distance or to time? Thus, if (6.41) seems evident to us, we have benefited from Galileo's investigations and Newton's development of calculus.

7.1.5 Pendulum Accelerometer

On plane flights, to estimate the power required for takeoff, I often estimate the plane's takeoff speed (its speed when it leaves the runway). To do so, I measure the plane's acceleration while on the runway. I tie my key chain to a string and let the key chain dangle downward. While the plane accelerates (while I feel pressed against the back of my seat), the string makes a nonzero angle θ with vertical (for example, with a window edge). On many flights, my homemade pendulum accelerometer (Figure 7.7a) reports that θ is roughly 10 degrees.

▶ *What's the plane's acceleration magnitude?*

As with most problems in Newtonian mechanics, a freebody diagram organizes the relevant information and our thinking. The pendulum bob (my key ring) interacts with the string and the earth. Thus, two forces act on it: the contact force from the string and the gravitational force from the earth (Figure 7.7b).

The gravitational force $\mathbf{F}_g$ has magnitude mg and points down. Compactly,

$$\mathbf{F}_g = m\mathbf{g}, \tag{7.25}$$

where **g** has magnitude g (the gravitational acceleration) and points down. The contact force $\mathbf{F}_{contact}$ points along the string toward my hand. Thus, its direction is known but not its magnitude (which is the unknown string tension).

◀ *Why is this freebody diagram so similar to the freebody diagram for the decelerating ice block (Section 7.1.2) and the accelerating bicycle (Section 7.1.3)?*

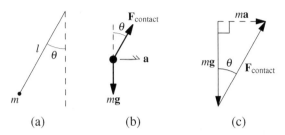

Figure 7.7 The Newtonian analysis of a pendulum accelerometer. (a) The pendulum accelerometer. (b) The freebody diagram of the bob, which experiences two
forces. (c) Adding these forces tip-to-tail to get the net force and, therefore, $m\mathbf{a}$.

In all three situations, the body experiences the same two forces – a gravitational force and a contact force – and has a rightward acceleration. And in all
three situations, the contact force performs the same two functions. Its vertical portion balances gravity. Its horizontal portion, the frictional force, causes
the rightward acceleration. Thus, the contact force, as the sum of the vertical
horizontal portions, points, in all three situations, upward and to the right.

Meanwhile, the situations differ in the direction of the velocity (opposite to
the acceleration for the ice block but along the acceleration for the bicycle and
the pendulum bob) and in the type of engine (none for the ice block and the bob
but you for the bicycle). Newton's laws and the freebody diagrams show you
the deep similarities under these surface differences.

Do we have enough equations to solve for the unknowns?

The unknowns are two: the magnitude of the contact force and the magnitude of
the acceleration. However, the number of equations seems to be only one, namely
Newton's second law connecting the net force on the bob to its acceleration.
Fortunately, the law is a vector equation, so it contains equations for magnitude
and direction. Its magnitude equation provides one equation. And specifying
a direction (in two dimensions) requires one number – for example, an angle.
Thus, the second law's direction equation provides our second equation. (In
three dimensions, specifying direction requires two angles, so the second law's
direction equation counts for two equations.) With two unknowns and two
equations, the problem should have a unique solution.

Emboldened by this check, let's find the solution. From the second law (4.6),

$$\underbrace{m\mathbf{g} + \mathbf{F}_{contact}}_{\mathbf{F}_{net}} = m\mathbf{a}. \qquad (7.26)$$

Thus, $m\mathbf{g}$ and $\mathbf{F}_{contact}$, drawn tip-to-tail to form their vector sum, produce
$m\mathbf{a}$ (Figure 7.7c). Graphically, the three vectors $\mathbf{F}_{contact}$, $m\mathbf{g}$, and $m\mathbf{a}$ form a
triangle. Because $m\mathbf{g}$ and $m\mathbf{a}$ are perpendicular, the triangle is a right triangle.

Its hypotenuse is the contact force, and its legs are the gravitational force $m\mathbf{g}$ and the scaled acceleration $m\mathbf{a}$. The definition of tangent gives us a/g:

$$\tan \theta \equiv \frac{\text{opposite leg}}{\text{adjacent leg}} = \frac{ma}{mg} = \frac{a}{g}. \tag{7.27}$$

With $\theta \approx 10°$, $\tan \theta$ and a/g are roughly 0.2. Thus, in answer to the triangle question (about the plane's acceleration magnitude),

$$a \approx 2\,\mathrm{m\,s^{-2}}. \tag{7.28}$$

On one flight where I made this measurement, the plane, a Boeing 747, accelerated for 40 seconds at this acceleration. Thus, its takeoff speed was roughly 80 meters per second (300 kilometers, or 180 miles, per hour). (To estimate how far this plane traveled on the runway, try Problem 7.2.)

7.1.6 How a Passive Force Adjusts Itself

Many contact forces are passive forces or contain a passive force – for example, the contact force on the accelerating bicycle (Section 7.1.3) or the tension force on the key chain in the pendulum accelerometer (Section 7.1.5). A passive force, as you learned in Section 1.2.2, adjusts itself in response to active forces. But *how* does a passive force adjust itself? How does it even know its correct value?

Our most familiar passive force is the normal force while we stand on the ground (Section 5.1). A similar example, where the adjustment mechanism is easier to picture, is the normal force on a soft, squishy rubber ball of mass m sitting on a rubber table. From the second law, $\mathbf{N}$ must balance the gravitational force, so $N = mg$. But how does the normal force know to have this magnitude?

Having practiced the second law with nonzero acceleration, you can use the second law to peer in slow motion into this mysterious adjustment process. To start, imagine holding the ball just above the table. Slowly lower it until it just touches the table and experiences zero normal force.

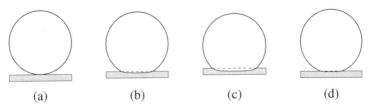

(a)   (b)   (c)   (d)

Figure 7.8 A rubber ball adjusting its normal force. (a) The ball barely touching the table just after being released ($N = 0$). (b) The ball after falling and compressing enough that the contact force balances gravity ($N = mg$). (c) The ball at maximum compression and zero velocity ($N > mg$). (d) The ball at a new minimum compression and zero velocity ($0 < N < mg$).

Now let the ball go (Figure 7.8a). With no force from your hand and no normal force (yet!), the only force on the ball is the gravitational force, mg downward. This force, which gives the ball an acceleration of g downward, sets the ball moving. The ball's subsequent motion can be divided into the four phases of spring motion: approaching equilibrium (phase 1), overshooting equilibrium and falling (phase 2), returning to equilibrium (phase 3), and overshooting equilibrium and rising (phase 4).

1. *Approaching equilibrium.* The ball speeds up because of the unbalanced gravitational force, falling ever faster. However, the table stands in its way. To make room for the ball, the ball and table compress, creating a contact interaction of increasing strength. Thus, the normal force on the ball, one side of this interaction, increases in magnitude. Eventually, N reaches mg. At this equilibrium point (Figure 7.8b), the net force on the ball and, therefore, its acceleration are zero. However, the ball is still falling.

2. *Overshooting equilibrium and falling.* To make room for the falling ball, the ball and table compress further, which increases N. With N now greater than mg, the net force on the ball and the resulting acceleration point upward – but the ball is still falling. The upward net force slows down (decelerates) the ball. Eventually, the ball stops. It has reached its lowest point: the point of maximum compression and maximum N (Figure 7.8c).

3. *Returning to equilibrium.* The net upward force (upward because $N > mg$ still) accelerates the ball upward. It rises ever faster, which reduces the compression of the ball and table and reduces N. Eventually, the ball reaches the equilibrium point again where $N = mg$, which makes the net force zero. Thus, the ball's acceleration is again zero, and it's rising at maximum speed.

4. *Overshooting equilibrium and rising.* The ball overshoots the equilibrium position, reducing the ball's and the table's compression and reducing N below mg. Now the net force and acceleration point downward (opposite to the ball's velocity). Eventually, the downward net force decelerates the ball to a stop (Figure 7.8d). The ball has reached its highest point.

This four-phase cycle – one period – then repeats itself. If the compression and uncompression were perfect, not turning any motion energy into heat or sound, the cycle would repeat itself with a fixed amplitude. In reality, these processes are imperfect. The rubber warms up, meaning that some motion energy has turned into heat energy; or the rubber's vibrations make sound, meaning that some motion energy has been turned into and carried away by sound waves. Thus, the amplitude of the oscillations decays, and the compression and the normal force converge to their equilibrium values – whereupon $N = mg$.

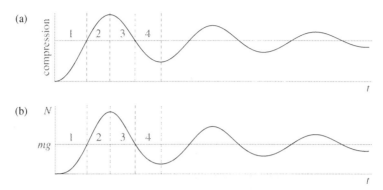

Figure 7.9 The ball's convergence to equilibrium (simulated). (a) The ball's compression versus time in the four phases. It eventually converges to the equilibrium compression of Figure 7.8b. (b) The normal-force magnitude N versus time in the four phases. It also eventually converges to the equilibrium value where $N = mg$.

The ball-and-table analog computer has computed the correct passive force! With this picture as the backstory, imagine again the squishy ball set onto the squishy table. If the table or the ball are squishy enough, the initial oscillations can be large enough (say, many millimeters) and their period long enough (say, a few hundred milliseconds) to notice. Although with stiff objects – a glass marble set on a wood table – the oscillations are too tiny and rapid to notice, the process that determines N remains the same. The compression of the object and the table, and the resulting N, oscillate around their respective equilibrium values (Figure 7.9). As energy leaves the system, these oscillations decay, and the compression converges to its equilibrium value: just enough for N to equal mg, whereupon it balances the gravitational force (the active force).

This slow-motion replay of how a passive force adjusts itself also illustrates the difference between two forces *balancing*, when they cancel in the net force, versus two forces *being equal and opposite*, when they are two sides of one interaction. Balanced forces, a matter for the second law, start unbalanced and stay unbalanced until the compression has adjusted itself. Equal and opposite forces, a matter for the third law, are instantly and always equal and opposite – at least in Newtonian mechanics. (Instantaneous adjustment violates special relativity, the successor to Newtonian mechanics, as you learn in Section 8.3.)

7.1.7 Forces on a Bouncing Ball

After analyzing your standing on a hill (Section 5.2), my sledding downhill at constant speed (Section 5.5), or your accelerating down a frictionless hill

(Section 7.1.4), you might conjecture that a body of mass m experiences a normal force with magnitude $N = mg \cos \theta$ (where θ is the hill's angle of inclination).

A simple test of this plausible, widely believed conjecture is applying it to a book resting on a table. A table is hill where $\theta = 0$ and $\cos 0 = 1$, so the conjecture correctly predicts that $N = mg$. With many examples in its favor, the conjecture must be true!

Or must it? As the great mathematics teacher George Pólya once said [17], it's the mark of a savage not to prove (or disprove) one's conjectures. To avoid renown as a savage, let's subject the $N = mg \cos \theta$ conjecture to further testing.

In particular, imagine a steel ball with mass m dropped from 1 meter high onto a steel table. The table and ball are steel to make the bounce nearly elastic: Most of the ball's energy comes back to it after the bounce, and it rebounds to almost 1 meter. Thus, energy loss is a minor effect, and we can concentrate instead on the forces. And to minimize their number, forget about air resistance.

What force(s) act on the ball while it's falling but before it hits the table?

In this first phase of its motion (Figure 7.10a), the ball touches no other bodies except the air. However, we are ignoring air resistance, so the ball participates in no short-range interactions and experiences no short-range forces. Meanwhile, it participates in only one long-range interaction, the usual gravitational interaction with the earth. Thus, it experiences only one long-range force: the gravitational force mg downward.

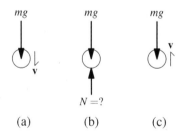

Figure 7.10 Freebody diagrams of the steel ball while falling, bouncing, and rising. (a) While falling. The ball experiences only gravity. (b) While bouncing, when the ball is instantaneously stationary. Now the ball experiences gravity and a normal force. (c) While rising. Now the ball, as in (a), experiences only a downward force, gravity – even though it's rising. Don't fall into the $\mathbf{F}_{net} \to m\mathbf{v}$ trap!

What about while the ball is rising after it hits and has left the table?

Because the ball is moving upward, our deep hardwired belief that $\mathbf{F}_{net} \to m\mathbf{v}$ (Section 4.1) tempts us to include an upward force – perhaps called $\mathbf{F}_{upward}$ or $\mathbf{F}_{motion}$ or, because the table sent the ball upward, $\mathbf{F}_{table}$. However, the preceding force analysis of the falling ball applies equally to the rising ball. Thus, in this third phase of its motion, the ball also experiences only one force: the gravitational force mg downward (Figure 7.10c).

▷ *During the bounce, what forces act on the ball at the instant when it's stationary?*

During the bounce, while the ball touches the table, the ball joins a new interaction: the contact interaction with the table. Thus, the ball experiences two forces: the usual gravitational force mg downward and the contact force N upward (Figure 7.10b). (Even if we weren't ignoring drag, it wouldn't matter at this instant anyway. The ball is stationary, so any air-resistance force is zero.)

▷ *At this instant, how large is N compared to mg?*

The ball is stationary, so perhaps the problem is one of statics (Chapter 5). In statics, the net force is zero. Therefore, the normal force, in order to balance the gravitational force, would have magnitude $N = mg$.

▷ *What's wrong with the preceding "it's statics, so $N = mg$" argument?*

The argument seems to prove too much. In milder bouncing, where a ball jiggled up and down as N eventually adjusted itself (Section 7.1.6), N almost never equaled mg. Furthermore, the argument feels dangerously wrong.

To see why, first rest a rock of mass m on your hand (itself resting on a table). The rock shouldn't be so large that it hurts your hand. In this familiar situation that definitely is statics, the normal force on the rock (from your hand) and on your hand (from the rock) have magnitude $N = mg$.

Now, in your imagination and with your other, free hand, raise the rock to a height of 1 meter and drop it onto your hand. If "it's statics, so $N = mg$" is valid, then, even while the rock bounces off your hand, the normal force on your hand (from the rock) still has magnitude mg. And mg didn't hurt. Yet, you still want to pull your hand away quickly, before the rock lands on it! Your gut can feel that the normal-force magnitude would be painfully larger than mg.

The flaw in the "it's statics" argument is the ancient confusion between velocity and acceleration and about which quantity is connected to (net) force. It's the struggle between our ancient intuition that $\mathbf{F}_{net} \to m\mathbf{v}$ and the modern, Newtonian view that $\mathbf{F}_{net} \to m\mathbf{a}$.

Statics, with its condition of zero net force, doesn't require or imply zero velocity; it implies zero acceleration. For you can give a body any velocity that you want – without changing its acceleration or the forces on it – simply by looking at it from a new inertial reference frame (Section 3.4). What matters for Newton's second law, and therefore for the normal force, is a body's acceleration. This acceleration, which is the same in all inertial frames, can be nonzero.

To see how far from zero the acceleration can be, let's estimate the ball's acceleration when it's stationary during the bounce (Figure 7.11b) by estimating an easier, related quantity: the ball's average acceleration over the whole bounce (while the ball touches the table). The clock starts when the ball, on its way down, first touches the table (Figure 7.11a). It stops when the ball, on its way up, loses contact with the table (Figure 7.11c).

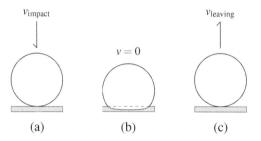

Figure 7.11 The steel ball during its bounce. (a) The clock starts: The ball has just touched the table on its way down. (b) The ball is stationary and at its maximum compression. (c) The clock stops: The ball has just left the table on its way up.

In symbols, and using the definition of average acceleration (6.22),

$$\underbrace{\mathbf{a} \sim \mathbf{a}_{avg}}_{\text{our estimate}} \equiv \frac{\Delta \mathbf{v}}{\Delta t}. \tag{7.29}$$

The numerator is

$$\Delta \mathbf{v} \equiv \underbrace{\left(\begin{array}{c} \text{ball's velocity when it} \\ \text{leaves the table} \end{array} \right)}_{\mathbf{v}_{\text{leaving}}} - \underbrace{\left(\begin{array}{c} \text{ball's velocity when it} \\ \text{first touches the table} \end{array} \right)}_{\mathbf{v}_{\text{impact}}}; \tag{7.30}$$

and the denominator is

$$\Delta t = \text{contact time (how long the clock runs).} \tag{7.31}$$

This estimate for **a** makes not only mathematical sense but also physical sense. To feel its physical meaning, multiply both sides of (7.29) by m:

$$m\mathbf{a} \sim \frac{m\Delta \mathbf{v}}{\Delta t}. \tag{7.32}$$

As an equation for magnitudes, (7.32) becomes

$$ma \sim m\frac{|\Delta \mathbf{v}|}{\Delta t}. \tag{7.33}$$

Now imagine that you've jumped from a table and are landing. The left side of the magnitude equation (7.33), namely ma, is also F_{net}, the magnitude of the net force on you when you land. On the right side are three factors, each reasonably placed given its effect on F_{net}.

1. $|\Delta\mathbf{v}|$. A higher $|\Delta\mathbf{v}|$ – for example, jumping from a higher table – should be more dangerous to your bones because of the larger net force on them. Thus, this factor should be, and is, in the numerator.

2. Δt. To protect your joints, you naturally bend your legs as you land. Doing so increases the contact time, Δt, and, being a safety measure, should decrease F_{net}. Thus, Δt should be, and is, in the denominator.

3. m. One drawback of leaving childhood (increasing m) is increased strain on the joints from the larger net forces on them. Thus, m should be, and is, in the numerator.

In summary, $\mathbf{a} \sim \Delta\mathbf{v}/\Delta t$ makes physical sense.

Estimating the ball's $\mathbf{a}$, then, requires estimating $\Delta\mathbf{v}$ using (7.30) – which requires estimating $\mathbf{v}_{impact}$ and $\mathbf{v}_{leaving}$ – and estimating the contact time Δt. As a rough estimate of $\mathbf{v}_{impact}$ for a 1-meter drop,

$$\mathbf{v}_{impact} \sim 1 \, \text{m s}^{-1} \text{ downward.} \tag{7.34}$$

(This estimate is low, by roughly a factor of 4, as you can calculate using the results of Problem 7.1. But this underestimate compensates for an upcoming underestimate of the contact time.) After the bounce, the ball has simply reversed direction (losing hardly any speed), so

$$\mathbf{v}_{leaving} \sim 1 \, \text{m s}^{-1} \text{ upward.} \tag{7.35}$$

Putting this velocity and $\mathbf{v}_{impact}$ from (7.34) into $\Delta\mathbf{v}$ from (7.30) gives

$$\Delta\mathbf{v} \sim 2 \, \text{m s}^{-1} \text{ upward.} \tag{7.36}$$

(If the collision isn't fully elastic, then $\Delta\mathbf{v}$ has a smaller magnitude, but that reduction won't change the surprising conclusion.)

Estimating the contact time (Δt) is trickier. It requires a rough slow-motion model of the bounce (Figure 7.11). First, the ball's bottom touches the table – and the clock starts. The ball's top keeps moving downward at v_{impact} while its bottom compresses (as does the table). The bottom, tired of being squeezed, shouts to the top, "Turn around! No more room! We've hit an unforgiving table!" When the message reaches the top, the top starts moving upward, uncompressing the ball (and table). Once the ball is completely uncompressed, the bottom loses contact with the table. The ball has left the table – and the clock stops.

This process takes roughly the time required for the "Turn around!" signal to traverse the ball. This signal is a sound wave. In steel, sound travels at roughly 5 kilometers per second. If the ball is conveniently 5 centimeters in diameter, then the contact time is only roughly 10 microseconds (μs):

$$\Delta t \sim \frac{5 \, \text{cm}}{5 \, \text{km s}^{-1}} = 10^{-5} \, \text{s} = 10 \, \mu\text{s}. \tag{7.37}$$

Turning the ball's velocity around, from $\mathbf{v}_{\text{impact}}$ to $\mathbf{v}_{\text{leaving}}$, in such a short time will mean a huge acceleration and net force.

For the curious. The estimate of Δt in (7.37) implicitly assumes that the ball's contact area doesn't change as the ball compresses. Accounting for that effect enlarges the contact time by roughly a factor of $(c_{\text{sound}}/v_{\text{impact}})^{1/5}$, where c_{sound} is the speed of sound in the ball. For a steel ball dropped from 1 meter, this correction is roughly a factor of 4. For simplicity, the following calculations use the uncorrected contact time.

To simplify the arithmetic in the calculation of acceleration, assume that the "Turn around" signal's journey from the bottom to the top of the ball constitutes one-half of the contact time (because the top, even after getting the signal, still has to stop and then speed up to v_{leaving}). Then, doubling the result in (7.37),

$$\Delta t \sim 2 \times 10^{-5} \text{ s}. \tag{7.38}$$

Using (7.29) for $\mathbf{a}$ in terms of $\Delta \mathbf{v}$ and Δt, (7.36) for $\Delta \mathbf{v}$, and (7.38) for Δt,

$$\mathbf{a} \sim \frac{\Delta \mathbf{v}}{\Delta t} \sim \frac{2 \text{ m s}^{-1} \text{ upward}}{2 \times 10^{-5} \text{ s}} = \underbrace{10^5 \text{ m s}^{-2}}_{10^4 g!} \text{ upward}. \tag{7.39}$$

This huge acceleration is, according to Newton's second law, caused by a comparably huge net force equal to $m\mathbf{a}$:

$$\mathbf{F}_{\text{net}} = m\mathbf{a} \sim 10^4 mg \text{ upward}. \tag{7.40}$$

The net force is the sum of the normal force $\mathbf{N}$ and the gravitational force, which contributes only $1mg$ downward. Because gravity is so tiny in comparison to the net force, the normal force is almost exactly the net force:

$$\mathbf{N} \approx \mathbf{F}_{\text{net}} \sim 10^4 mg \text{ upward}. \tag{7.41}$$

This analysis has three key lessons. First, the normal force's magnitude N need not be mg or even the more general $mg \cos \theta$. When the ball first touches the table, N is zero. When the ball is stationary during the bounce and maximally compressed, N is roughly $10^4 mg$. During the rest of the bounce, N takes all values in between. Sorry, conjecture!

Second, the steel ball bouncing from a table has an important similarity to the squishy ball placed on a table (Section 7.1.6): Both balls are acted on by a passive force, the normal force, that changes as the ball oscillates up and down, going through the four phases of spring-like motion. Springs are everywhere.

On the squishy ball, N starts at zero and oscillates around and converges to its equilibrium value of mg – as the ball goes through the four phases of spring-like motion repeatedly. On the bouncing steel ball, N also starts at zero, when the ball first touches the table (Figure 7.12). The ball then goes through the four phases of spring-like motion but completes only one cycle. (1) The steel ball, falling, quickly reaches the equilibrium position, where $N = mg$. (2) It falls past the equilibrium position and quickly reaches maximum compression, producing a huge upward normal force. (3) This force accelerates the ball upward, returning it quickly to the equilibrium position. (4) The ball rises past the equilibrium position and uncompresses rapidly. When its compression is gone, $N = 0$, and the ball is about to lose contact with the table. It's then rising at its maximum speed, roughly equal to v_{impact}. With this speed, the steel ball continues upward and rises above the table – instead of returning to phase 1 as the squishy ball would. Thus, N never gets to converge to its equilibrium value (mg).

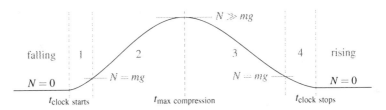

Figure 7.12 The normal-force magnitude N versus time (simulated). (Falling) The ball is falling but in the air ($N = 0$). (1) The ball has touched the table and, as it falls, is compressing toward the equilibrium compression (where $N = mg$). (2) The ball continues to fall and compress until it reaches maximum compression. (3) The ball rises and uncompresses until it reaches equilibrium again. (4) The ball overshoots equilibrium toward zero compression. (Rising) The ball has left the table.

As the third lesson, Newton's second law confirms your gut feeling: When a rock is about to fall onto your hand, move your hand away quickly. Because the bounce from a squishy hand lasts longer than the bounce from a hard steel table, the normal force on your hand is significantly less than $10^4 mg$. This reduction, however, isn't good news. The bounce from your hand takes longer partly because of the time for your hand bones to bend and break. Newton's laws speak with one voice: Please move your hand!

Can N ever be negative?

I almost answered "yes," showing how deep the confusion of vectors and scalars can run. However, N is a magnitude and can never be negative. But a related question has an interesting answer: Can the normal force itself ($\mathbf{N}$ rather than N) point out of a body (rather than only into the body)? It can, and it means that the normal force is attractive (rather than repulsive). As one example, glue a coin to your palm and face your palm downward. Then the normal force on the coin, which is one side of its interaction with your hand, balances gravity and points upward (Figure 7.13a) – which means that it points out of the coin.

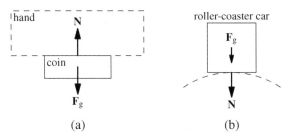

(a) (b)

Figure 7.13 Freebody diagrams with an attractive normal force. (a) A coin glued to your palm. The normal force belongs to the attractive interaction between the coin and your hand and points out of the coin. Thus, following the conventions in Section 2.1 for drawing freebody diagram, its *tail* is placed at the contact surface. (b) A roller-coaster car zooming over the top of a hill. At fast enough speeds, the track has to grip the roller coaster – which means providing an attractive (downward!) normal force.

Another example is a roller-coaster car zooming over the top of a hill. If it might ever move fast enough, faster than the roller coaster that you analyze in Problem 7.19b, then the track needs a grip mechanism that holds the car on the track. The grip makes the contact interaction between the track and the car attractive and the normal force on the car point out of the car (Figure 7.13b).

7.2 Composite Bodies

The safety advice of Section 7.1.7 concludes our introduction to acceleration for motion in one dimension. Before we brave the complexity of motion in two dimensions (Section 7.3), we'll discuss an issue that I have skated around and is implicit in most of the preceding examples: composite bodies. For a single particle, or for a body treated as a single particle, Newton's second law is (4.6).

But what happens to the second law when the body is composite – composed of more than one particle – as is every real-life body? For example, what if the body is a box plus bumblebees (Section 2.3), a cyclist plus bicycle (Section 5.4), or a car with tires and an engine?

Newton's second law (4.6) links three quantities: (1) the body's mass, (2) the net force on the body, and (3) the body's acceleration. In revising the law to apply to a composite body, each quantity changes in notation and interpretation.

1. *Mass.* As you might expect, this quantity becomes the composite body's mass. This mass is often labeled M, to distinguish it from m, but is also often labeled m_{total} or even just m.

2. *Net force.* Taken literally, this quantity is the sum of *all* forces on *all* particles in the composite body. That calculation seems daunting. For example, consider the composite body of you and the bicycle (Section 5.4). How do you find the force that you exert on the bicycle and the force that the bicycle exerts on you? They can be found, but the process is elaborate and requires several freebody diagrams. Without knowing these two forces, finding $\mathbf{F}_{net}$ and using Newton's second law look impossible.

   To the rescue comes Newton's third law. The contact force on you from the bicycle and the contact force on the bicycle from you are two sides of one interaction, so they are equal and opposite. Thus, in the grand sum of all forces on all particles, these two forces balance: They add to zero. It's magic: To find $\mathbf{F}_{net}$, you don't need either force because their sum, all that matters for $\mathbf{F}_{net}$, is zero!

   This magic happens for every *internal* interaction: for every interaction between two parts of the composite body. Because each internal interaction contributes nothing to the net force, the net force can be calculated simply from the body's *external* interactions. Each such interaction contributes one force to the net-force sum. This sum, the net *external* force, is often labeled $\mathbf{F}_{net}^{ext}$. With that notation, the magical conclusion is that

$$\mathbf{F}_{net} = \mathbf{F}_{net}^{ext}. \tag{7.42}$$

   The net external force is also often labeled $\mathbf{F}_{net}$ – with no indication of the new interpretation. Then you just have to remind yourself to consider only external forces – as mentioned in commandment 2 about freebody diagrams (p. 28) and as we did when making the freebody diagram for sledding (Section 5.5.1), where we ignored the internal forces between me and the sled.

3. *Acceleration.* This quantity, like net force, also needs closer specification. In the composite bicycle body, is **a** your acceleration or the bicycle's? These accelerations can be different: Imagine standing up in the pedals while riding along at constant velocity. The bicycle has no acceleration, but you do (unless you stand up at constant speed, which is hard to do).

   It turns out – meaning, it's true, but I won't prove it – that the relevant acceleration is neither yours nor the bicycle's. Rather, it's the acceleration of the composite body's center of mass. This acceleration is usually labeled $\mathbf{a}_{CM}$. But sometimes it's just **a**. Then you have to remind yourself that **a** stands for the center-of-mass acceleration. But how do you calculate this acceleration?

For a composite body, the *location* of its center of mass is the weighted average of the individual particle locations (Section 2.1). Similarly, the *velocity* of its center of mass is the weighted average of the individual particle velocities. And, continuing the pattern, the *acceleration* of its center of mass is the weighted average of the individual particle accelerations:

$$\mathbf{a}_{CM} = \frac{m_1\mathbf{a}_1 + m_2\mathbf{a}_2 + \cdots + m_n\mathbf{a}_n}{m_1 + m_2 + \cdots + m_n}. \qquad (7.43)$$

With these notation and interpretation changes, the generalized second law is

Newton's Second-and-a-Half Law. $\quad \dfrac{1}{M}\mathbf{F}_{net}^{ext} \to \mathbf{a}_{CM}. \quad (7.44)$

Because this revised version dispenses with internal forces thanks to incorporating the third law, I call it Newton's second-and-a-half law.

In words, the net *external* force on a body, modulated in its effect by the body's mass, accelerates the body's center of mass. Thus, Newton's laws describe only how forces change the motion *of* the composite body's center of mass. How the composite body moves *about*, or relative to, its center of mass – in other words, how it rotates and how that rotation changes – is left undetermined. (That determination requires a new physical concept, torque, and the rotational version of Newton's laws – ideas touched upon in Section 8.2.)

7.3 Two-Dimensional Motion

A tempting misconception in Newtonian mechanics, with roots partly in the ancient confusion between velocity and acceleration, is thinking that the net force on a body lies along the same line as the body's motion – that is, that the net force lies along the velocity. This misconception is tempting because, in one-dimensional motion, the net force and velocity do lie on a common line. For example, a cyclist moving in a straight line, whether speeding up or slowing down, has a net force along the same line.

However, this simple relation is too simple. For example, a thrown rock follows a parabolic path (neglecting air resistance), but the net force on it and its acceleration point directly downward – a direction never aligned with the velocity. Force and velocity have no necessary connection.

The necessary connection is between force and acceleration: (Net) force causes acceleration. And, as you saw in Section 6.3, acceleration in two (or more) dimensions is qualitatively different than in one dimension. In two dimensions, a body's acceleration can have a portion perpendicular to the body's velocity. Thus, the net force can too – again unlinking net force and velocity.

When I remember this point, the misconception becomes less tempting. Because studying one-dimensional motion for too long makes it ever more tempting, we now turn to two-dimensional motion.

7.3.1 Force on an Orbiting Satellite

As our first extended example of motion (velocity) and force pointing in different directions, Figure 7.14 shows a satellite at several positions in its orbit.

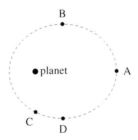

Figure 7.14 A satellite at several positions in its orbit.

▶ *At each position, what forces act on the satellite?*

The satellite's only interaction is its gravitational interaction with the planet, so only one force acts on the satellite: gravity (Figure 7.15a). It always points toward the planet (the satellite's partner in the interaction). Its magnitude is inversely proportional to the square of the distance from the planet (Section 1.3.1), so it's the largest at C, the smallest at A, and in between at B and D.

▶ *Do the forces depend on whether the satellite orbits clock- or counterclockwise?*

No! The gravitational force on the satellite depends only on its location relative to the planet. The satellite's velocity, except indirectly by changing the satellite's location, has no effect on the force.

▶ *Do the net force and the velocity ever lie along a common line (as they always would in one-dimensional motion)? What about the acceleration and velocity: Do they ever lie on a common line?*

The gravitational force, which is also the net force, always points directly at the planet. But the satellite never moves directly toward the planet. Thus, the net force and the velocity never lie along a common line. Similarly, the acceleration, which has the same direction as the net force and also points directly at the planet, never lies along the velocity. Two-dimensional motion is qualitatively different from one-dimensional motion.

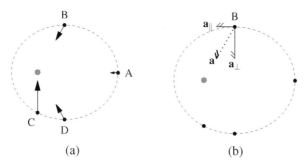

Figure 7.15 The orbiting satellite. (a) The forces on the satellite at various points. At each point, the only force, gravity, points directly at the planet holding the satellite in orbit. (b) The satellite's acceleration at point B split into parallel and perpendicular portions.

Now assume that the satellite orbits counterclockwise. At point B, is it speeding up (accelerating) or slowing down (decelerating)?

To decide, use a key idea about acceleration: that it can be broken into perpendicular and parallel portions (Section 6.6), each reflecting one aspect of the vector nature of velocity. The perpendicular portion $a_\perp$ arises from the velocity changing *direction*. The parallel portion $a_\parallel$ arises from the velocity changing *magnitude* – from the satellite speeding up (accelerating) or slowing down (decelerating). When the satellite speeds up, $a_\parallel$ points forward (along v). When the satellite slows down, $a_\parallel$ points backward (opposite to v).

Throughout the satellite's orbit, its acceleration a points directly at the planet. At point B, the planet is ahead (though not directly ahead) of the satellite. Thus, the $a_\parallel$ portion of a points along v, directly ahead of the satellite (Figure 7.15b). Therefore, the satellite is speeding up. This conclusion makes physical sense: At point B, the satellite is falling toward, though fortunately not directly toward, the planet; thus, it should be speeding up.

7.3.2 Circular Motion

The next example of two-dimensional motion is another problem that I learned in J. W. Warren's works [25, pp. 35–37]. The problem is to describe the forces on a cornering car – a car going around a turn (Figure 7.16a). In particular, the car moves clockwise along a circular road at constant speed; furthermore, there is no wind (the air is still, relative to the road). Compared to the satellite of Section 7.3.1, this car follows a simpler trajectory but experiences more forces.

▶ *On Figure 7.16a (a top view), draw the following forces on the car: (1) the net force $\mathbf{F}_{net}$, (2) the friction force $\mathbf{f}$ exerted by the road on the car, and (3) labeled arrow(s) representing any other force(s) acting on the car. Ignore forces or their portions lying outside the horizontal plane.*

Did you draw your forces? If not, go back and do it! Then you'll learn much more from the following discussion. Figure 7.16b shows a typical diagram, whose forces may show similarities to your set of forces.

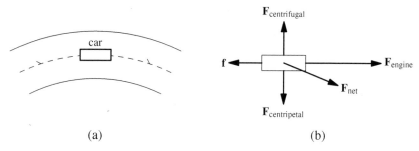

(a)   (b)

Figure 7.16 (a) A car going clockwise around a circular path at constant speed (top view). (b) A typical but incorrect freebody diagram (top view).

▶ *In this typical diagram, how many mistakes can you find?*

Finding all the errors means solving the problem. And experts solve problems by working forward from known to unknown quantities (in contrast to novices, who work backward) [12]. So, if you have become a Newtonian expert, great. If not, fake it till you make it. In either case, work forward from what you know.

You know the car's motion, including how it changes. Thus, the problem is of type I (inferring): From the car's motion, infer the forces acting on it. In particular, from the car's acceleration $\mathbf{a}$, use the second law to infer the net force that must have caused $\mathbf{a}$:

$$\mathbf{F}_{net} = m\mathbf{a}. \qquad (7.45)$$

The car's acceleration is familiar: The car moves in a circle at constant speed. In physics jargon, the car is in uniform circular motion, which you studied in Section 6.3. The results were twofold. First, and less important here: $a = v^2/R$, where v is the car's speed, and R is the circle's radius. Second, and more important here: Because the car's speed is constant, $a_\parallel$ is zero, leaving only a perpendicular portion, so $\mathbf{a}$ points directly inward.

From (7.45), a directly inward acceleration requires a directly inward net force. That observation reveals the first problem in the typical diagram: $\mathbf{F}_{\text{net}}$ doesn't point in the same direction as $\mathbf{a}$ – a violation of Newton's second law.

This incorrect net force, oriented mostly forward and slightly downward, is common. I find myself tempted by it too. Its orientation reveals the many-headed hydra of the force-of-motion trap: our strong, intuitive, and incorrect conviction that (net) force causes velocity.

My own internal incorrect reasoning, too fast for me to catch in detail, runs roughly as follows: "The car's velocity points straight ahead of the car, so the net force should also be straight ahead. [Whoops: I've just fallen into the force-of-motion trap!] But that net force would move the car straight ahead, whereas the car needs to move in a circle by turning slightly to the right as it moves clockwise along the road. Thus, I'll twist the net force slightly to the right, so that it points mostly forward and slightly downward." If you also catch yourself reasoning similarly, stop in the first sentence, where net force is assumed to cause velocity, and return to Newton's second law: Net force causes acceleration. Also, remind the intuitive part, which so loves $\mathbf{F}_{\text{net}} \to m\mathbf{v}$, that it should chew instead on quantity of motion, introduced inSection 7.1.2, and that force changes quantity of motion.

The second problem in the typical diagram is revealed by the meaning of "net" in "net force." All the actual forces on the car have to add up to the net force. In the typical diagram, these forces – $\mathbf{f}$, $\mathbf{F}_{\text{centrifugal}}$, and $\mathbf{F}_{\text{centripetal}}$ – do not add up correctly, neither to the net force as drawn nor to the correct net force that points directly inward. So, these forces also cannot be right.

> *What should these forces be?*

To dynamite the fog of forces, return to the first principle: interaction (Chapter 1). Each force on the car comes from one of its interactions. The easiest interactions to find are usually long range, of which there is usually only one, the gravitational interaction with the earth, as is true here. To find the short-range (contact) interactions, consider the bodies touching the car. Here, they are the road and the air. (Although there is no wind, meaning that the air isn't moving relative to the road, the air hasn't disappeared and moves relative to the car.)

Thus, the car participates in three interactions, one long range and two short range, so it experiences three forces: gravity (from the gravitational interaction with the earth), the contact force (from the contact interaction with the road), and air drag (from the contact interaction with the air).

Because the diagram, which is a top view, shows motion and forces only in the horizontal plane, the gravitational force will not appear on the diagram. However, air drag, $\mathbf{F}_{\text{drag}}$, is purely horizontal and appears in its entirety. The contact force, $\mathbf{F}_{\text{contact}}$, has a vertical portion $\mathbf{N}$ (the normal force), which balances the gravitational force, but the only part of $\mathbf{F}_{\text{contact}}$ that appears on the diagram is its horizontal portion $\mathbf{f}$, the frictional force of the road acting on the tires.

In summary, there are only two forces in the horizontal plane: $\mathbf{F}_{\text{drag}}$ and $\mathbf{f}$. Thus, none of $\mathbf{F}_{\text{centrifugal}}$, $\mathbf{F}_{\text{centripetal}}$, or $\mathbf{F}_{\text{engine}}$ belongs on the diagram.

◀ *What's wrong with those forces?*

$\mathbf{F}_{\text{centrifugal}}$ is a familiar interloper (Section 1.5.1). It's not an actual, physical force. Neither is $\mathbf{F}_{\text{centripetal}}$ (Section 1.5.2). Thus, neither force belongs on the diagram.

The most subtle interloper is $\mathbf{F}_{\text{engine}}$. It seems like a physical force: What use is an engine that exerts no force? However, the force that the engine exerts is an *internal* force, one side of the interaction between the engine and the rest of the car. The interaction's other side, the force of the rest of the car on the engine, is similarly an internal force. By the third law, these two internal forces are equal and opposite and, together, contribute nothing to the net force on the car.

In summary, only the external forces acting on the car affect the car's (center-of-mass) acceleration – an example of Newton's second-and-a-half law (7.44). Thus, $\mathbf{F}_{\text{engine}}$ doesn't belong on the car's freebody diagram. (Alternatively, you could include $\mathbf{F}_{\text{engine}}$, shorthand for $\mathbf{F}_{\text{engine on rest of car}}$. But then you also have to include its third-law counterpart $\mathbf{F}_{\text{rest of car on engine}}$, which would balance it in the net-force sum.) The car gets its acceleration from the road and from the air, the external bodies, not from the engine.

◀ *If the engine isn't accelerating the car, what's it doing?*

Just as you do when you accelerate a bicycle (Section 7.1.3), the engine helps create the correct conditions at the point of contact between the road and the tire – conditions that produce the necessary static-friction force $\mathbf{f}$.

The other creators of these conditions are the driver and the steering wheel. By turning the steering wheel, the driver turns the front wheels. Their orientation relative to the back wheels along with the distance between the front and back axles are the two factors that together determine the radius of curvature of the car's path (Figure 7.17).

(a)   (b)

Figure 7.17 Turning the front wheels to determine the radius of curvature of the car's path. (a) Front wheels turned slightly: large $r_{\text{curvature}}$. (b) Front wheels turned sharply: small $r_{\text{curvature}}$.

To keep the car on the road, the driver orients the front wheels so that this radius of curvature equals the road's radius of curvature R. Meanwhile, R and the car's speed v determine **a** (which is v^2/R directly inward) and, by the second law, the required net force on the car. Thus, the wheels' orientation helps determine the inward contribution to $\mathbf{F}_{net}$ needed from the frictional force **f**.

Like an honorable witness who tells the whole truth and nothing but the truth, the analysis now contains all the correct forces – $\mathbf{F}_{drag}$ and **f** – and no others. The final step is to ensure that the net force is the sum of the actual forces:

$$\mathbf{F}_{net} = \mathbf{F}_{drag} + \mathbf{f}. \tag{7.46}$$

In direction, you know two of these three forces: $\mathbf{F}_{net}$, because you know the acceleration direction; and $\mathbf{F}_{drag}$, because you know the motion direction. To draw these forces, you also need their magnitudes. For this rough drawing, just choose reasonable relative magnitudes. Then use the geometry of vector addition to find the frictional force **f** (Figure 7.18a). The forward portion of the resulting **f** balances the air drag $\mathbf{F}_{drag}$, leaving its inward portion to be the net force and to produce the car's inward acceleration. Now you can make the corrected freebody diagram (Figure 7.18b).

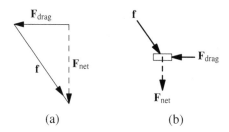

(a)   (b)

Figure 7.18 Making the corrected freebody diagram. (a) Finding the frictional force **f**. It and the drag force $\mathbf{F}_{drag}$ add tip-to-tail to make the net force $\mathbf{F}_{net}$ (dashed because it's not an actual force). (b) Including this **f** on the car's corrected freebody diagram (top view). $\mathbf{F}_{drag}$ and **f** are repulsive contact forces, so they are drawn with their tips on the edge of the car.

A note on tips and tails. The corrected freebody diagram follows the useful convention that a contact force should be drawn with its tip or tail on the edge of the body at the force's point of application. Thus, the diagram implies that **f** is applied at the edge of the car's footprint on the road, roughly on the line extending the rear axle. This incorrect implication results from projecting three-dimensional drawings onto two dimensions.

In fact, the point of application is located such that the line of action of the contact force (of the sum of **f** and **N**) passes through the car's center of mass.

Usually, this point lies inside the car's footprint. However, placing **f**'s tip inside the footprint would incorrectly imply that **f** is a long-range or body force.

In this game of choose your poison, I kept the contact-force convention. This problem would anyway be resolved in a three-dimensional drawing, where the tip of **f** (or $\mathbf{F}_{contact}$) would touch the bottom of the car. Thus, the three-dimensional diagram would show the vector lying outside the car and touching its surface.

7.3.3 Centripetal Force: Clearing Up Misconceptions

In making a freebody diagram, you may have been tempted to label an inward force as the centripetal force $\mathbf{F}_{centripetal}$ – for example, on the cornering car (Figure 7.16b). If you've ever been tempted, read on for a health warning.

▷ *What kind of force is the centripetal force?*

The simplest answer comes from the adjective's Latin meaning (Section 1.5.2). The Latin root "centri" means "center," and "petal" comes from the Latin "petere" meaning "seeking," so centripetal means "center seeking." A centripetal force is a force toward a center, usually directly toward the center.

However, as you learned in Section 1.5.2, a centripetal force isn't a fifth kind of force additional to the four fundamental forces of Section 1.2.1. Rather, "centripetal" merely redescribes an actual, physical force – almost always a gravitational or an electromagnetic force – that points directly toward a center.

◁ *Should I use the "centripetal" label?*

Because it labels merely an existing force, it won't help you draw freebody diagrams. And it can cause confusion, especially if you look for *the* centripetal force. Here are examples of traps for the unwary.

Return to the cornering car of Section 7.3.2 but first without air drag. This car moves in a vacuum at constant speed on the circular path. Vacuum or not, the net force still points directly toward the center. Without air drag, the only force on the car in the horizontal plane is static friction **f** (the horizontal portion of the contact force of the road). Meanwhile, the vertical portion of the contact force of the road balances the gravitational force. Thus, the static friction equals the net force and likewise points directly toward the center.

But look what can go wrong when you use the centripetal force. You start with the correct idea that, on a body in uniform (constant-speed) circular motion, some force points toward the center of the circle. You then draw such a force and label it $\mathbf{F}_{centripetal}$ (Figure 7.19a). But the label leaves unclear the status of that force and thereby invites misconceptions.

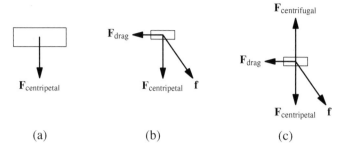

Figure 7.19 The problems of the centripetal force. (a) Labeling the net force as $\mathbf{F}_{\text{centripetal}}$. (b) Including the two real forces. Now it's easy to forget that the centrifugal force isn't a real force and is only the sum of the two real forces. Forgetting that point makes the net force look incorrect. (c) Fixing the net force by adding a bogus centrifugal force. Using the centripetal force has led to a bigger problem!

Is that inward force the net force? That resolution is plausible (and correct): The net force does point directly inward. However, that choice merely postpones the question of what actual, physical force the centripetal force labels. For the net force isn't a physical force. It's just a mental, or mathematical, construct created to help us state and apply Newton's second law (which can then say simply that net force produces acceleration).

Instead, is that inward force the frictional force $\mathbf{f}$? You might find this (correct) supposition hard to believe – I sometimes do – because of the conviction that friction "opposes motion." This ball of confusion usually remains untangled because it produces a correct freebody diagram with exactly one force pointing directly inward (never mind what to call it!).

The ball of confusion gets tangled beyond repair when air drag is included (as in Section 7.3.2). Now, drag points backward, and static friction points inward and forward: Only their sum, the net force, points directly inward. Thus, no actual, physical force points directly toward the center. However, because uniform circular motion requires a (net) force directly toward the center – that is, a (net) centripetal force – you might start looking for it and include it on the freebody diagram (Figure 7.19b). (For this freebody diagram, I have drawn $\mathbf{F}_{\text{drag}}$ and $\mathbf{f}$ with their tails at the car's center of mass, as if they were body forces like gravity rather than contact forces: Anyone careful enough to place these forces at the edge of the car is unlikely to fall into the centripetal-force trap.)

Nothing good comes of that choice. Either the inward-pointing arrow stands unrecognized for what it is, the net force – costing you a reminder of how net force causes acceleration and a reminder to determine what physical forces produce the net force – or this inward-pointing arrow gets imagined as an actual, physical force. But it seems strange that an actual force would point directly toward the center. You might therefore attempt to balance the alleged centripetal force and thereby nullify its effect by also including a *centrifugal* force pointing directly outward (Figure 7.19c). One misconception leads to a second.

To avoid them, remember two points. First, the "centripetal force" isn't itself a physical force (that is, it isn't a new, fifth kind of force beyond the four kinds in Section 1.2.1, of which the most important are gravitational and electromagnetic forces). It is just a label for a physical force that points (directly) toward the center. Second, the net force is not a physical force either. Thus, even when the net force points directly toward the center, as it does in uniform (constant-speed) circular motion, it should not be labeled as a centripetal force.

As further protection, neither ask about nor look for "*the* centripetal force." Instead, if you must talk about centripetal force, ask about "*a* centripetal force." The definite-article formulation ("the") implies exactly one centripetal force, which goads you into desperately sticking the label onto any force that points toward the center, even the net force. For it sounds ungrammatical to answer "Where is *the* centripetal force?" with "Nowhere." In contrast, the indefinite-article formulation ("a") allows the possibility of no centripetal force. For it sounds (mostly) fine to answer "Where is *a* centripetal force?" with "Nowhere."

Thus, much work is required to avoid problems when using the centripetal force. The simplest solution is never to use it. I never do except when correcting freebody diagrams (sometimes my own, alas). Instead, remember the first principle: that every force is one side of an interaction. Thus, find all (external) interactions in which a body participates, short range and long range; draw one force for each interaction; and draw no other forces!

7.4 Weight

Weight, like two-dimensional motion, is a subtopic that also invites many confusions. Unlike with two-dimensional motion, where the confusions are mostly innate, with weight the confusions are mostly self-inflicted and easier to resolve. To evoke and start to resolve the confusions, we first go into orbit (Section 7.4.1). Then we return to earth and calculate weight while swinging (Section 7.4.2) and while skiing downhill (Section 7.4.3). These calculations raise the question of how we feel weight (Section 7.4.4). Finally, with a clear understanding of weight, we solve the bumblebee problem (Section 7.4.5). As a bonus, the analyses provide frequent practice with the second and third laws, with their application to two-dimensional motion, and with freebody diagrams.

7.4.1 Orbiting

In orbit, the confusions around weight become most evident.

▷ *In low-earth orbit – for example, as an astronaut in a space shuttle or space station – are you weightless?*

The answer depends on how you define weight. If you define it as the magnitude of the gravitational force, a popular choice, then the astronaut isn't even close to weightless. For the earth's gravity is almost as strong at a low-earth-orbit altitude as it is on the ground. For example, at an altitude of 200 kilometers, the earth's gravity has almost 94 percent of its ground-level strength (Problem 1.5).

However, if you define weight as what a spring (or weighing) scale displays, then you – floating about in the spacecraft and exerting no contact force on any support or scale – are weightless. Before I reconcile these conflicting answers, you may wonder about floating itself (on which the second answer depended).

◁ *Why or how does an astronaut float?*

This deep question calls for a Newtonian analysis of an astronaut and spacecraft in orbit. As the sole force on you the astronaut (Figure 7.20), the gravitational force alone gives you your inward acceleration of magnitude g_{orbit} – where g_{orbit} is roughly 9.4 meters per second squared, reflecting the slightly weaker gravity in low-earth orbit. As long as your orbital speed v satisfies

$$\frac{v^2}{R_{\mathrm{orbit}}} = g_{\mathrm{orbit}}, \tag{7.47}$$

where R_{orbit} is your orbital radius (your distance from the center of the earth, so your altitude above the ground plus the radius of the earth), you move in a circular orbit at constant speed – the uniform circular motion of Section 6.3.

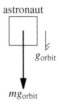

astronaut

g_{orbit}

mg_{orbit}

Figure 7.20 Your freebody diagram in orbit. You participate in only one interaction, which is gravitational, and therefore experience only one force: gravity.

Meanwhile, the gravitational force on the spacecraft gives the spacecraft the same inward acceleration and same circular-orbit speed as you have. Thus, you and the spacecraft orbit in tandem, meaning that you float around in the spacecraft with no contact force on you.

This conclusion changes temporarily if the spacecraft changes to a different orbit by blasting its engines. The spacecraft now experiences an additional force, the contact force from its interaction with the exploding fuel. This force modifies the spacecraft's acceleration and, eventually, its orbit. While the orbit changes, you and the spacecraft no longer orbit in tandem. Thus, the spacecraft eventually touches and pushes you – exerting a contact force on you. This new force changes your acceleration and velocity until your orbit matches the spacecraft's new orbit. Once you and the spacecraft orbit in tandem, the contact force on you vanishes, and you float – making you again weightless according to the spring-scale definition of weight.

How then do you reconcile the two definitions of weight?

The fundamental problem arises not from the mere existence of two different definitions of weight but rather from mixing them – which happens easily. Most people, including me, use the spring-scale definition implicitly: "The astronaut is weightless!" Meanwhile, most people – and I was among them once – use the gravitational-force definition explicitly. It's taught in most textbooks, where many freebody diagrams use mg and W (weight) interchangeably: "The astronaut's weight is mg."

Unfortunately, the two conflicting conclusions – weightless versus weight is mg – are most easily reconciled by concluding that, on a body in low-earth orbit, the gravitational force is zero (thus, that g_{orbit} is zero). The falsity of the conclusion may seem like a mere hiccup, easily fixed by reminding ourselves that gravity in low-earth orbit is nearly as strong as on the ground. However, this reminder treats merely the symptom while masking the underlying confusion: mixing the two definitions of weight. Their mixture is especially problematic because the forces pointed to – spring versus gravitational forces – represent different kinds of fundamental forces (electromagnetic and gravitational, respectively) and have different ranges (long range and short range, respectively).

The proper treatment is to pick one definition and to use it explicitly and implicitly. Either definition could work. However, changing an explicit belief is easier than changing an implicit belief – as I remember whenever I encounter in myself remnants of the implicit, pre-Newtonian belief that force causes velocity. This reason points toward changing our explicit definition to match our implicit, spring-scale definition. Furthermore, creating a new name (weight) for an existing name (the gravitational-force magnitude) fosters confusion because it implies that the two names refer to different forces. Thus, I recommend and use the spring-scale definition: Weight is what a spring scale displays.

What, exactly, does a spring scale display?

A spring scale displays the magnitude of the normal force on its top surface. The way that it works conceptually (an actual spring scale, such as a bathroom scale, has many refinements) is that the normal force depresses the top surface and compresses a spring underneath it. The compression distance x and the spring's stiffness – its spring constant k in (1.18) – together determine the normal force's magnitude N (as kx). Confusingly, N is then displayed as a mass m using $m = N/g$ (based on a canonical value of g at the earth's surface) – for a metric scale, which gives its reading in grams or kilograms (a unit of mass).

If the scale displays pounds, the force-to-mass conversion is superfluous: In the British gravitational system and in both US engineering units systems (foot–pound–second and inch–pound–second), the pound is officially a unit of force. Unfortunately, this surprising choice, justifiable three centuries ago when it was implicitly made, foments widespread and deeply rooted confusions about weight, mass, and force because it carries into our time the inevitable confusions of the era of its origin, when mass and force were first studied. My solution is to pretend that the pound is a unit of mass – what I learned in grade school during America's anemic attempt to metricize and what most people think it is anyway. I then spring to safety by immediately converting these unofficial pounds to kilograms, at approximately 2.2 unofficial pounds per kilogram, and continue the calculation in metric – a system of units that won't surely confuse me.

With the spring-scale definition of weight, an astronaut floating around in a space shuttle, who exerts no contact forces but experiences gravity at almost full strength, is weightless.

Orbiting is admittedly unusual even for astronauts, who spend most of their days on the ground. Thus, the spring-scale definition of weight may seem like overkill. First, it seems to solve a mostly hypothetical problem yet adds the actual complexity of solving for a passive, contact force. Second, in many earthbound situations, the normal force that a body exerts on the scale anyway equals (in magnitude) the gravitational force on the body – for example, stand peacefully ($\mathbf{a} = 0$) on a bathroom scale. By the second law, the normal force on you has magnitude mg (Section 5.1). Therefore, by the third law, the normal force that you exert on the scale also has magnitude mg. Thus, why not just use the simpler, gravitational-force definition of weight?

The answer is that the equality depends on the condition of zero (vertical) acceleration. When a body's acceleration has a nonzero vertical portion – say, while bouncing (Section 7.1.7), swinging (as we'll analyze in Section 7.4.2), or in orbit – the gravitational force and the normal force differ in magnitude and the definition of weight matters. The spring-scale definition's conceptual clarity more than compensates for its extra calculational complexity.

7.4.2 Swinging

Returning to earth with the spring-scale definition of weight and heeding the caution about nonzero vertical acceleration, we next calculate my weight in such a situation: while standing on a playground swing (a pendulum) that is swinging back and forth (Figure 7.21). The chains holding up the swing have a length l of approximately 2 meters. At the bottom of the swing's arc, the swing's seat, my feet, and the thin spring scale just below them travel at, say, 4 meters per second (14 kilometers, or 9 miles, per hour).

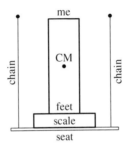

Figure 7.21 Me standing on a playground swing (rear view).

> *When the swing is at the bottom of its arc, what's my weight? That is, what force does the scale display in comparison to its regular reading of mg?*

Finding the weight requires finding the normal force on me – hence, drawing my freebody diagram. My only long-range interaction is the gravitational interaction. My only short-range interaction is the contact interaction with the scale. (For this calculation, we'll neglect air resistance, but you can consider its effect by doing Problem 7.15.) Thus, only two forces act on me: the downward gravitational force and the upward normal force.

   The last step in constructing the freebody diagram is to include on it my acceleration – but the acceleration of which part of me? Should it be the acceleration of my feet (because they touch the scale) or maybe of my head (because it's where Newton's laws are being used)? My different parts have different speeds and might also have different accelerations. I'm a composite body.

   The answer comes from Newton's second-and-a-half law (7.44), repeated here with m as my mass:

$$\frac{1}{m}\mathbf{F}^{\text{ext}}_{\text{net}} \rightarrow \mathbf{a}_{\text{CM}}. \tag{7.48}$$

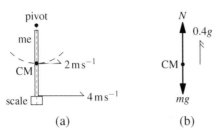

Figure 7.22 Analyzing swinging. (a) The geometry and motion of me standing on the playground swing (side view). The dashed vertical line is the chain. My center of mass (CM) moves in a circle whose radius is roughly 1 meter: The scale is thin, my CM is roughly at my midpoint, and, in my dreams, I am 2 meters tall (like the chain). (b) My freebody diagram. The normal force adjusts itself to give my CM an acceleration of 0.4g upward.

The relevant acceleration – the one caused by the net (external) force – is the acceleration of my center of mass (CM), which moves in a circle centered on the pivot (Figure 7.22a). In circular motion, the acceleration has an inward, perpendicular component

$$|\mathbf{a}_\perp| = \frac{v_{CM}^2}{r}, \tag{7.49}$$

where r is the radius of the circle in which my center of mass moves, and v_{CM} is the speed of my center of mass. My center of mass lies roughly halfway from the pivot to my feet. Thus,

$$r \approx \frac{1}{2} \times l = 1 \text{ m}, \tag{7.50}$$

and

$$v_{CM} \approx \frac{1}{2} \times \underbrace{4 \text{ m s}^{-1}}_{v_{feet}} = 2 \text{ m s}^{-1}. \tag{7.51}$$

The parallel portion of the acceleration, by its direction, says whether my center of mass is speeding up or slowing down. At the bottom of the swing arc, v_{CM} is a maximum, so my center of mass is doing neither ($dv_{CM}/dt = 0$). Thus, the parallel portion is zero, leaving a purely upward acceleration – as the pendulum bob of Section 6.5 has at point C, the bottom of its arc.

With the estimated values for r in (7.50) and for v_{CM} in (7.51),

$$a_{CM} = \frac{v_{CM}^2}{r} \approx \frac{(2 \text{ m s}^{-1})^2}{1 \text{ m}} = 4 \text{ m s}^{-2} = 0.4g. \tag{7.52}$$

Now the freebody diagram is mostly complete (Figure 7.22b); the only unknown is N, the magnitude of the normal force. As a passive force, it quickly adjusts itself to the value that keeps my center of mass moving in its circle. That value is determined by Newton's second-and-a-half law (7.48). This law is a vector equation, but only the vertical or y portions matter here (the horizontal force portions and accelerations are all zero, so they produce the old news that $0 = 0$). With up as the positive y direction, the y component of (7.48) gives

$$\underbrace{N - mg}_{F^{net}_y} \rightarrow ma_{CM}. \tag{7.53}$$

Using $a_{CM} \approx 0.4g$ from (7.52) and solving for N,

$$N \approx 1.4mg. \tag{7.54}$$

As a consequence of the third law, N is also the magnitude of the normal force that I exert on the scale, so it is my weight (answering the triangle question). As expected given my upward acceleration, $N > mg$: I feel much heavier at the bottom of the swing's arc than I do when standing peacefully on the ground.

7.4.3 Skiing

You too can change your weight by accelerating. In our next example, you stand on a triangular wedge (Figure 7.23) as you and the wedge ski down a hill inclined at $\theta = 30°$. (Actually, you must crouch or squat to avoid falling over backward: After you have studied Section 8.2, try Problem 8.2.) A scale is conveniently built into the top surface of the wedge. In order to concentrate on calculating weight and the associated contact forces, ignore air drag and friction between the wedge and the hill. (But friction between the wedge and you is important, as you analyze in Problem 7.20.)

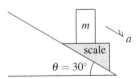

Figure 7.23 Zooming down a frictionless hill on a wedge. A weighing scale is built into the wedge's top surface.

▶ *What's your weight? That is, what force magnitude does the scale display?*

The scale displays the magnitude of the normal force acting on its top surface. Equivalently, it displays the magnitude of the normal force on you – these two normal forces, we learn from Newton's third law, are equal in magnitude. Thus, to find the normal force on the scale, we need to make just your freebody diagram (which is simpler than the freebody diagram of the wedge because you participate in fewer contact interactions than the wedge does).

Your freebody diagram gets one force from each of your two interactions: from the gravitational interaction with the earth and from the contact interaction with the wedge or, equivalently, with the scale built into the wedge's top surface. Thus, you experience two forces: the gravitational force mg downward and a contact force $\mathbf{F}_{\text{scale}}$. The magnitude of $\mathbf{F}_{\text{scale}}$'s upward portion – that is, the magnitude of the normal force $\mathbf{N}$ – is the scale reading.

Thus, N is the ultimate goal, and an intermediate goal is $\mathbf{F}_{\text{scale}}$, which is determined by your acceleration $\mathbf{a}$. Whoops – by saying that acceleration determines force, I just committed the mind-projection fallacy (Section 5.5.1)! Rather, $\mathbf{F}_{\text{scale}}$ and the gravitational force together, as the net force, determine $\mathbf{a}$.

Fortunately, we can determine $\mathbf{a}$ without first finding $\mathbf{F}_{\text{scale}}$. For you and the wedge are sliding together down a frictionless plane – the analysis done in Section 7.1.4. Its result is that $\mathbf{a} = g \sin \theta$ downhill.

Knowing this $\mathbf{a}$, we could laboriously deduce the $\mathbf{F}_{\text{scale}}$ that, along with the gravitational force, produces this $\mathbf{a}$. Fortunately, we have already done it, when you zoomed down the perfectly oiled hill (Section 7.1.4). Your acceleration there was $g \sin \theta$ down the hill, as it is here. There, you experienced a downward gravitational force mg and a contact force – just as you do here. Thus, the contact force in the two situations must also be the same. There, it was $mg \cos \theta$ perpendicular to the hill – which must also be $\mathbf{F}_{\text{scale}}$ here. (The wedge, thanks to the X-ray vision provided by Newton's laws, becomes invisible.)

The scale displays N, the vertical component of $\mathbf{F}_{\text{scale}}$ (Figure 7.24). Because $\mathbf{F}_{\text{scale}}$ is tilted by θ relative to the vertical,

$$N = \underbrace{F_{\text{scale}}}_{mg \cos \theta} \cos \theta = mg \cos^2 \theta. \tag{7.55}$$

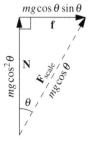

Figure 7.24 Resolving the contact force $\mathbf{F}_{\text{scale}}$ into its perpendicular portion $\mathbf{N}$, whose magnitude N the scale displays, and its parallel portion $\mathbf{f}$.

With $\theta = 30°$ and $\cos \theta = \sqrt{3}/2$, we can answer the triangle question:

$$N = \frac{3}{4}mg. \tag{7.56}$$

Skiing down this plane on a wedge, you weigh only three-fourths of your usual weight. Skiing is a recipe for instantly losing weight – though not for losing mass (what most of us are interested in).

7.4.4 Feeling Weight

The distinction between weight and gravitational force raises the question of how we even sense or feel weight. The answer turns on the distinction between long- and short-range forces acting on a body. Imagine yourself again standing on level ground (Section 5.1). One force on you is "the" long-range, gravitational force – with "the" in quotes because gravity acts on every tiny morsel of you. Thus, zillions of tiny gravitational forces are distributed throughout your body (Figure 7.25). The other force is the short-range, contact force of the ground acting on the soles of your feet.

Figure 7.25 A more accurate freebody diagram of you standing on the ground. The gravitational force, rather than being represented as a single force (and arrow), is split into many tiny forces. (The normal force, **N**, could also be split. It would become many tiny forces along your bottom surface.)

A surface or contact force and a volume or body force (here, the gravitational force), just by their very names, necessarily act at different places in you. This difference produces stresses within you and compresses your tissues and bones (you built a simple model in Problem 5.8). This compression is what you feel and to which you are accustomed.

Differences from the accustomed compressions are strongly noticed. When you zoom over a hill in a roller-coaster car (what you analyze in Problem 7.16), your stomach feels as if it were rising into your chest. Although the gravitational force on you hasn't changed just by your ascending a few meters to the start of the roller-coaster track, the contact force acting on your underside has changed.

Therefore, the internal stresses are different from what you are used to – and, therefore, so are the internal compressions and stretches (technically, the strains). These differences create the feel and excitement of a roller-coaster ride.

Similarly but over a longer term, when astronauts live for a long time with low weight – for example, in a space station – their bones do not grow normally. The bones do not feel the usual stresses that they feel on earth, stresses that would stimulate bone growth and repair.

A plant also feels the difference between volume (long-range) and surface (short-range) forces. As I learned from J. W. Warren's discussion [25, p. 21]:

> Even a tiny seed is stressed by the noncoincident opposing forces of gravity and the upthrust of soil, which is why the root is stimulated to grow downwards when the seed germinates.

To cement the idea that weight shouldn't be defined as gravitational force, consider the following situation with no gravitational force and only a normal force (Figure 7.26): You stand in an elevator in deepest space, far from any appreciable gravitational fields, and the elevator accelerates upward with a constant acceleration of magnitude g, the gravitational acceleration on earth. (Defining an upward direction is difficult, maybe impossible, without gravitational fields, but that valid point doesn't vitiate the upcoming analysis.)

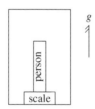

Figure 7.26 Standing on a scale in an accelerating elevator. The elevator's acceleration is g upward.

▶ *What weight does a scale under your feet read?*

You accelerate identically to the elevator, at g upward, so the net force on you must be mg upward. This net force is supplied by the normal force – the sole force acting on you. By the third law, the normal force on the (top surface of the) scale is mg downward. So, the scale displays mg, as it would on earth.

▶ *Does it feel, throughout your body, like standing on the earth? Or is the similarity only in the scale reading?*

It indeed feels like standing on the earth throughout your body (not just where you touch the scale, at the soles of your feet). That is, the internal forces on each morsel of you are the same as if you stood on the earth. To see why, compare corresponding forces in the two situations. When you stand on the earth, each morsel, with mass Δm, experiences two slightly unbalanced compressive forces (Figure 7.27a). Their vector sum must equal $g\Delta m$ upward – in order to balance the tiny downward gravitational force on the morsel and prevent it from accelerating. When you stand in the accelerating deep-space elevator, the vector sum must also equal $g\Delta m$ upward – in order to give the morsel its acceleration of g upward (Figure 7.27b).

Thus, in the elevator as on the earth, the top and bottom forces on corresponding morsels differ by an identical amount ($g\Delta m$). Furthermore, at the contact surface – whether the elevator floor or the ground – the corresponding bottom forces are identical. Therefore, the internal, compressive forces – the top and bottom forces – are identical in the two situations. The accelerating elevator has artificial earth gravity.

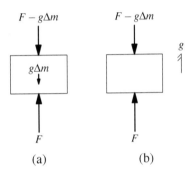

$$F - g\Delta m \qquad F - g\Delta m$$

$$g\Delta m$$

$$g$$

$$F \qquad F$$

$$(a) \qquad (b)$$

Figure 7.27 Comparing standing on the earth to standing in the accelerating elevator. (a) The freebody diagram of a morsel of you when you stand on the earth. The internal, compressive forces together balance gravity, leaving the morsel with zero acceleration. (b) The freebody diagram of a morsel when you stand in the elevator. The compressive forces are the same as in (a); however, without gravity to balance them, they give the morsel an upward acceleration of magnitude g.

◀ *Is this method used to create artificial gravity for astronauts on long trips (to recreate the internal stresses that keep their bones growing normally)?*

It is, with a modification needed to resolve the following problem. The accelerating elevator – starting from rest and after accelerating for, say, 10 hours – travels over 6000 kilometers, comparable to an orbital radius around the earth, and moves at roughly 360 kilometers *per second*. Reaching this speed, far higher than even the earth's orbital speed around the sun, requires a massive amount of fuel. For any useful duration, straight-line acceleration is impractical.

Yet, the fundamental idea – using acceleration to mimic gravity – is sound. Its cheaper implementation is to spin a spacecraft or space station on its axis (Figure 7.28). If the edge, where the astronauts live, rotates at a constant speed v

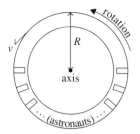

Figure 7.28 A space station that is rotating in order to provide artificial gravity.

and $v = \sqrt{gR}$, where R is the distance from the axis to the edge, then, from (6.57), the astronauts' inward acceleration has magnitude g:

$$a = \frac{v^2}{R} = \frac{gR}{R} = g. \tag{7.57}$$

The spacecraft or space station has artificial gravity! Artificial gravity illustrates one extreme of feeling forces, where a single external force is felt (through its consequences internally).

Are there, at the other extreme, forces that cannot be felt?

Yes! Imagine a long-range force – also called a body or volume force (Section 1.2.3) – applied *uniformly* throughout a body's volume. "Uniformly" means that, to each of the body's morsels, the force contributes the same acceleration (the same in magnitude and direction). Such a force changes the velocity of all morsels identically. Therefore, it does not affect the motion of the morsels relative to each other. So, it affects neither the body's internal structure – the compressions and stretches – nor its internal forces. The force cannot be felt.

An example of such a volume force is the gravitational force on a body in a uniform gravitational field, one where the strength and direction of gravity are the same everywhere. This gravitational force cannot be felt. For this reason, Einstein realized that gravity is a funny sort of force and reimagined it not as a force but rather as a change in the geometry – the curvature – of space and time. (That magnificent story is introduced in Section 8.3.2.)

Although no gravitational field is perfectly uniform, most can be approximated as uniform far enough from the gravitational source and over a small-enough region. (The small-region requirement is why Einstein needed to build his theory of gravity upon *differential* geometry, which examines the geometry of space and time over tiny regions and then stitches the analyses together – similarly to how differential calculus assembles curves out of tiny straight regions.)

The earth's gravitational field is nearly uniform near (or above) the surface, 6000 kilometers from the center of the earth. In particular, it's nearly uniform throughout a region small compared to 6000 kilometers (for example, throughout a person). Thus, we don't feel the earth's gravity! Instead, we feel the compressions (and stretches) produced by internal forces – themselves the result of the ground preventing us from falling freely.

Also reasonably uniform is the Sun's gravitational field at the earth's orbital distance, even over a region as large as the earth. Because the sun's gravity is the only significant force acting on the earth, the earth, in its orbit around the sun, feels no gravity. It is weightless!

7.4.5 Bumblebees

Having assembled the conceptual pieces – the spring-scale definition of weight (Section 7.4.1) and Newton's second-and-a-half law for composite bodies (Section 7.2) – we can now solve the bumblebee problem introduced in Section 2.3. A box of bees sits on a scale but is 10 grams over the weight limit. To clarify the colloquial language (necessary because grams are a unit of mass rather than weight): The normal force N (Figure 2.10) is 10 grams times g, or 0.1 newtons, too high. Might the bees help reduce the weight? In physics language, can N be reduced by convincing the bees to change their motion?

▶ *What happens to N if, when the bees are sleeping peacefully on the floor of the box, the truck driver whacks the box, waking up the bees who then take off?*

When the bees are resting, the composite body of box and bees has zero acceleration. In particular, its center of mass has zero acceleration. Thus, by the second-and-a-half law (7.44), the net external force on the body must be zero. So, the normal force $\mathbf{N}$ and the gravitational force $\mathbf{F}_g$ must balance: $N = F_g$. Here, F_g is the magnitude of the gravitational force on the composite body, so

$$N = F_g = (M + m)g, \tag{7.58}$$

where M is the mass of the box, and m is the mass of all the bees.

During the takeoff, the bees have, on average, a nonzero upward velocity; otherwise, they would never get airborne. "On average" means their center of mass. Thus, their center-of-mass velocity $\mathbf{v}_{CM}^{bees}$ points upward. Because $\mathbf{v}_{CM}^{bees}$ started at zero, when the bees rested on the box floor, $\mathbf{a}_{CM}^{bees}$ must also point upward (for some time in the takeoff) – otherwise $\mathbf{v}_{CM}^{bees}$ would never point upward.

Meanwhile, the composite body's other part, the box, has zero acceleration. The composite body's center-of-mass acceleration $\mathbf{a}_{CM}$ is the weighted average (7.43) of its two parts' accelerations (weighted by their respective masses). Thus, $\mathbf{a}_{CM}$, as the weighted average of a zero acceleration (for the box) and an upward acceleration (for the bees), also points upward.

To accelerate the center of mass upward, N must become larger than F_g (which isn't affected by the takeoff). Making the bees take off has *increased* the weight – our sad answer to the triangle question. But despair not! Although the weight moved in the wrong direction, the analysis has established the surprising but welcome news that the weight depends on the bees' motion.

▶ *Can the bees help reduce the weight?*

If their center of mass could just accelerate downward, N would be less than F_g. As the first step toward creating that acceleration, let the bees finish taking off and reach their respective cruising altitudes. Then $\mathbf{a}_{CM}$ is zero, and $N = F_g$ again (as when the bees were sleeping on the box floor). Now pump nontoxic sleeping gas into the box. The bees fall asleep midflight and go into free gravitational motion (free fall) toward the box floor. Their acceleration and the composite body's center-of-mass acceleration now point downward. N is less than F_g!

▶ *By how much does N fall?*

In free gravitational motion, the bees are acted on only by the gravitational force (neglecting air resistance), which no one feels anyway (Section 7.4.4). So, they are weightless, and their usual weight mg disappears from the scale reading given in (7.58). Thus, during free gravitational motion,

$$N = \underbrace{(M + m)g}_{F_g} - mg = Mg. \tag{7.59}$$

The scale displays just the (usual) weight of the box. (This conclusion, which I supported using a slightly glib argument about weightlessness, can be justified: Try Problem 7.24.) When the bees (free) fall, so does the weight.

The original scale reading was 10 grams over the limit, meaning an excess normal force of 10 grams times g. Thus, as long as the bees' mass is more than 10 grams, the problem is solved. And so we conclude our main course: Newton's second law in the general case of accelerating, composite bodies. The next and final chapter, which discusses what comes next after Newton's laws, offers dessert. Read on if you're hungry!

7.5 Problems

7.1 How fast is the toast of Section 7.1.1 moving when it lands (if you cannot catch it)? Ignore air resistance (and the toast's rotation). Work out the impact speed for a general table height h and then for $h \approx 0.6$ meters.

7.2 How far did the Boeing 747 of Section 7.1.5 travel on the runway while speeding up for takeoff?

7.3   At $t = 0$ and using a tennis racket, you launch a stone directly upward with speed $v_0 = 20$ meters per second (72 kilometers, or 44 miles, per hour). This launch speed, by the way, requires that the racket, at the point where and at the moment when it hits the stone, move upward at $v_0/2 = 10$ meters per second. In answering the following questions, ignore air resistance (and use $g = 10$ meters per second squared).

   a.   When does the stone reach its highest point? (When is its speed zero?)

   b.   During its upward journey, does the stone experience a force in the direction of motion (as did the falling stone of Section 7.1.1)?

   c.   During its upward journey, what are the stone's average speed, average velocity, and average acceleration?

   d.   During its upward journey, how far does the stone travel?

   e.   How long after its launch does the stone return to its launch height?

   f.   Over the round-trip journey, what are the stone's average speed, average velocity, and average acceleration?

7.4   In an (ideal) Atwood machine with two equal masses ($m_1 = m_2$), neither mass accelerates (the result of Section 5.6.3). In this problem, you find the acceleration in general, when m_1 and m_2 could be unequal (Figure 5.27a).

   a.   Make a freebody diagram for each mass.

   b.   Use the two diagrams to find T, the string tension, and a_z, the vertical component of m_1's acceleration.

   c.   Check (i) that your results for T and a_z are correct in the easy case $m_1 = m_2$ and (ii) that the sign of a_z makes physical sense when $m_1 > m_2$ and when $m_2 > m_1$.

7.5   After computing the kicked ice block's average speed (7.12), I stated that a constant **a** implies a steady change in speed only for one-dimensional motion and only if the body doesn't reverse direction. Illustrate these caveats as follows:

   a.   Give an example of two-dimensional motion where **a** is constant but the speed does not change steadily.

   b.   Give an example of one-dimensional motion where **a** is constant but the speed does not change steadily.

7.6   On the bicycle of Section 7.1.3, find F_{contact} in terms of m, a, and g.

7.7   In Problem 5.8, you modeled a person as three stacked blocks, in order to find the forces acting on and within a person standing on the ground. In this problem, you redo the analysis for a person freely falling toward the earth. Thus, imagine that you have just walked off a diving board and are in the air on the way to hitting the water (and ignore air resistance).

a. Draw freebody diagrams for the earth and for each of the three blocks used to model you.

b. How do the internal forces differ compared to when you stood on the ground in Problem 5.8?

c. How does the motion of the earth differ compared to when you stood on the ground?

7.8 The conclusion in Section 7.1.4, that $\mathbf{a}_\perp = 0$ when you slide down a frictionless ramp, depended on two conditions: (1) You move only along the hill, and (2) the hill is a straight line. To show that the $\mathbf{a}_\perp = 0$ conclusion needs the second condition, create a counterexample. That is, design a hill where, even though the first condition is satisfied, $\mathbf{a}_\perp$ isn't zero.

7.9 You are skiing down a frictionless hill holding a pendulum accelerometer (a mass attached to a string). The hill makes an angle θ with the horizontal. Assuming no air resistance, what angle does the string make with the vertical (once the mass has stopped oscillating)? What angle does the string make with the hill?

7.10 Still assuming a counterclockwise orbit for the satellite of Section 7.3.1:

a. At point A, what's $\mathbf{a}_\parallel$? And what's happening to the satellite's speed there? (Is it increasing, decreasing, or constant?)

b. At points C and D, in what direction does $\mathbf{a}_\parallel$ point? What's happening to the satellite's speed at these points?

How do your answers change if the satellite orbits clockwise?

7.11 Hoping to liven up the turn, the driver of Section 7.3.2 depresses the accelerator, and the car speeds up as it travels along the circular road. When its speed has doubled, the driver returns the car to constant-speed motion. Modify the horizontal-plane freebody diagram and the vector sum (Figure 7.18) to show the forces and net force at the following times:

a. just after the car starts speeding up (so, before its speed has increased noticeably), and

b. after its speed has doubled, and it has finished speeding up.

7.12 Figure 7.29 shows a common, incorrect freebody diagram of a cyclist and bicycle, considered as one body, the "cyclist," riding along a circular path at constant speed. The diagram is a front view showing the forces, or their portions, in the vertical plane. The cyclist, bicycling out of the page toward you, leans inward in order to travel in a circle.

a. What's correct in the diagram?

b. What are the errors in the diagram?

c. Draw a correct freebody diagram.

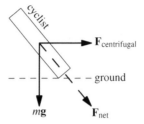

Figure 7.29 An incorrect freebody diagram of a cornering cyclist (really, of the person and the bicycle considered as a composite body).

7.13 For the car driving counterclockwise at constant speed around the elliptical track of Figure 6.8: At each of the points A, B, C, and D, draw the car's freebody diagram showing the net force $\mathbf{F}_{net}$, the frictional force $\mathbf{f}$ from the road, and any other forces (or portions thereof) in the horizontal plane. How do the diagrams change if the car drives clockwise (at constant speed)?

7.14 Deducing the period of a pendulum, even at small amplitudes, seems to require calculus because the pendulum's speed varies as it moves along its arc. But Christiaan Huygens (1629–1695), called "the most ingenious watchmaker of all time" [5, p. 79] by the great physicist Arnold Sommerfeld, found a calculus-free way by analyzing the motion of a conical pendulum: a pendulum moving in a horizontal circle (Figure 7.30). Now you can follow in his footsteps.

Figure 7.30 A conical pendulum (side view from above the orbital plane). The bob orbits counterclockwise in a horizontal circle (meaning that θ stays constant).

The pendulum consists of a string of length l and a bob (of mass m) in uniform circular motion. In the snapshot of Figure 7.30, the bob is moving into the page.

a. Find the bob's orbital speed v in terms of g, l, and (trigonometric functions of) θ. *Hint*: Make a freebody diagram of the bob, or repurpose the analysis of the pendulum accelerometer (Section 7.1.5).

b. Find the bob's orbital period T in terms of g, l, and θ.

c. For small θ, you can approximate $\sin\theta$ by θ and $\cos\theta$ by 1. In this limit, what's the period T?

Projecting the conical pendulum's two-dimensional motion onto a vertical screen – for example, the page – produces one-dimensional pendulum motion (for small θ). Thus, the period T, which you have calculated without using any calculus, is the period of one-dimensional pendulum motion!

7.15 How does air resistance affect my weight at the bottom of the arc while swinging, given in (7.54)? For the sake of comparison, assume that my feet still move at 4 meters per second. Is my weight still $1.4mg$, is it smaller, or is it larger?

7.16 You (a point of mass m) are sitting in a roller-coaster car that has just reached the peak of the curved track and is moving at 5 meters per second (Figure 7.31). At the peak, the track is 10 meters above the ground and has a radius of curvature of 5 meters.

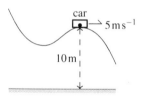

Figure 7.31 You sitting in a roller-coaster car that has just reached the peak.

a. What's your weight at the peak, expressed in terms of your usual weight mg? (You're a point moving exactly along the track.)
b. Roughly how fast could you be moving at the peak without needing a seat belt to hold you in the roller-coaster car?

7.17 What's your weight while standing on the wedge of Section 7.4.3 but waiting to ski? (The wedge is, for the moment, attached to the hill.)

7.18 As a function of m, g, and θ (the hill's inclination), what weight (normal force) does a scale under a skier's feet display (Figure 7.32)? This scale is built into the ski boots. Thus, in contrast to skiing on the wedge of Section 7.4.3, where the scale was horizontal, this scale is angled parallel to the hill. Do the analysis first without air resistance (for example, just after starting to ski). Then redo it with the air-resistance force produced at the skier's terminal speed (that is, after skiing for a long time).

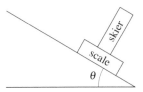

Figure 7.32 Skiing down a frictionless hill on a scale built into the ski boots.

7.19 A rider – who, like me, has mass $m = 60$ kilograms but, unlike me, isn't scared stiff of roller coasters – starts in a roller-coaster car at the highest point of a roller-coaster track (Figure 7.33). The rider, a curious physics student, has placed a bathroom scale under his or her bottom.

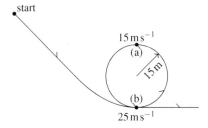

Figure 7.33 The looped roller-coaster track for Problem 7.19. Its (a) and (b) labels refer to the problem's subparts.

   a. At the top of the circular loop, which has radius 15 meters (approximately 50 feet), the rider is upside-down and traveling at 15 meters per second (54 kilometers, or approximately 34 miles, per hour). What force (in newtons) does the scale display? Explain (i) why this positive scale reading means that the rider needs no seat belt to stay in the car, and (ii) how a rider on whom no upward force acts stays in the car (the surprising situation mentioned, but not explained, in Problem 2.2).

   b. At the bottom of the loop, the rider is traveling faster, at 25 meters per second. What force does the scale display there? (The high reading is why a circular-loop design isn't used for modern roller coasters.)

7.20 When you ski downhill on the wedge of Section 7.4.3, what's the physical effect of the horizontal portion of $\mathbf{F}_{scale}$? In other words, what bad consequence would result if that portion vanished?

7.21 For the dropped steel ball of Section 7.1.7, imagine that the table contains an inbuilt weighing scale. Sketch a graph of the scale reading (in units of mg) versus time while the ball is touching the table.

7.22 Figure 7.34 shows the path of a solid rubber ball thrown forward. The path, familiar from Section 2.2.2, includes slightly inelastic bounces (where the ball loses energy and speed). At points A (just before the first bounce), B (at the first peak), and C (during the second bounce), draw the forces acting on the ball at that point. Include any air resistance, and use the lengths of the force arrows to show how the various forces rank in magnitude (the weakest force should have the shortest arrow, etc.). To avoid the complications of spin, assume that the ball–ground contact is frictionless. (This problem is adapted from one used by J. W. Warren [25, pp. 33–34].)

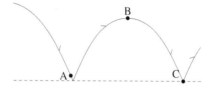

Figure 7.34 The path of a bouncing ball. Point A is just before the first bounce.

7.23 One day, I set food onto a digital kitchen scale, set the scale and food on the floor of our apartment building's elevator, and noted how the weight (as usual, expressed confusingly in units of mass) varied as the elevator traveled between the first and sixth floors (in British dialect, between the ground and fifth floors). Use my rough graph of weight versus time (Figure 7.35) to answer the following questions.

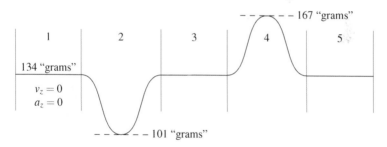

Figure 7.35 Weight of the food (expressed in mass units) versus time.

a. In regions 2–5 of the graph, give the sign (+, –, or 0) of v_z (the upward component of the elevator's velocity) and of a_z (the upward component of the elevator's acceleration). Thus, did I use the elevator to go up from the first to the sixth floor or to go down?

b. What was the elevator's approximate maximum upward acceleration in units of g?

c. What was the elevator's approximate maximum downward acceleration in units of g?

7.24 Use Newton's second-and-a-half law (7.44) to show that, while the bees of Section 7.4.5 are in free gravitational motion, the scale displays Mg (as I claimed in (7.59) but did not prove).

7.25 Sled dogs are accelerating three connected sleds in a straight line across frictionless ice (Figure 7.36). (The dogs' claws dig into the ice, providing the needed static friction.) In terms of m_1, m_2, m_3, and a, find (a) the string tension T_3, (b) the string tension T_2, and (c) the string tension T_1. *Hint:* Use composite bodies.

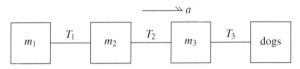

Figure 7.36 Three connected sleds pulled by sled dogs.

8

What Comes Next

The preceding chapters introduced the essential ideas of Newton's laws of motion. Yet, motion houses a vast store of mysteries. Surely, the story of Newton's laws cannot end so quickly. This final chapter touches upon ideas key to further study of mechanics: noninertial reference frames (Section 8.1), including how to use the dangerous centrifugal force safely; torque and rotation (Section 8.2); and going beyond Newton's laws (Section 8.3).

8.1 Noninertial Reference Frames

So far in this book, noninertial reference frames – ones that fail the first-law test (Section 3.2) – have been forbidden. For if a frame fails this test, then Newton's second law, the heart of mechanics, cannot be used in that frame.

Yet, you might need such a frame anyway. It might be the earth frame, which fails the test for many reasons: The earth rotates on its axis, the earth orbits the sun, the sun itself (and thus the solar system) orbits around the center of the galaxy, the galaxy orbits within our galactic cluster, and so on at ever-larger spatial scales.

Even if these reasons stopped at motion within the galaxy, no one wants to perform mechanics calculations in the galaxy frame (even assuming it to be an inertial frame) and then have to account for the sun's orbit in the galaxy, for the earth's orbit around the sun, and, finally, for the earth's rotation. If Galileo had needed that much rigmarole just to analyze cannonballs dropped from the leaning tower of Pisa [3], the scientific revolution would never have started. There must be a way to do mechanics directly in the earth frame, noninertial though it may be.

To reveal the questions that such frames create, go to the earth's equator and find a tall tower. From the top of the tower, drop a rock (Figure 8.1). If the earth were an inertial frame, the rock would land at the bottom of the tower. But where does it land on the actual earth? To keep the problem as simple as possible, forget about air resistance, about the earth's orbit around the sun, about the sun's orbit in the galaxy, and about all motion at larger scales. The only reference-frame motion is the rotation of the earth on its axis.

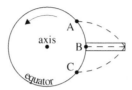

Figure 8.1 A tower at the equator of on a rotating earth (view from above north pole). A rock is dropped from the top of the tower. Which path does it follow: Does it land at point A, B, or C?

Where does the rock land: ahead of the tower (point A), at the bottom of the tower (point B), or behind the tower (point C)?

To decide, remember that Newton's second law is valid in an inertial reference frame. Thus, you can calculate the rock's motion in a base inertial frame and translate the result into the noninertial reference frame. Here, the base inertial frame sits above the north pole and observes the earth rotating counterclockwise.

In that frame, the ball starts not with zero velocity (which is its velocity in the earth frame) but rather with a velocity that points forward (along the direction of rotation). Its speed is the speed of the top of the tower. Meanwhile, the bottom of the tower also has a forward velocity. However, the bottom, being closer than the top to the rotation axis, moves slightly more slowly than the top moves. Thus, as the ball falls, it creeps ahead of the bottom of the tower and lands ahead of the tower (landing at point A). This result holds even after including a subtle countervailing effect: Because of the earth's rotation, the direction of gravity also rotates slightly as the ball falls.

For a 100-meter tower, the speed difference and the countervailing effect together produce a deflection that turns out to be only roughly 3 centimeters – which can matter for precise calculations. Furthermore, for longer falls, the deflection grows. In falling from a 100-kilometer tower (if one could be built), the stone would deflect by 0.9 kilometers.

If you didn't know better – if you didn't know that the rotating earth is a noninertial frame – and you applied Newton's laws anyway, you would be puzzled. The only force on the rock is the downward force of gravity, and the rock starts from rest. Thus, how can the rock's fall deviate from purely vertical?

The short answer is that using Newton's second law in a noninertial reference frame requires adding fictitious forces. These forces are not actual, physical forces: They are not among the four types of interactions in nature (Section 1.2.1). Nor do they obey Newton's third law. They are neither one side of an interaction, nor do they have a counterpart force. However, like the gravitational force, to which they bear an uncanny resemblance that Einstein used to create general relativity (as we'll discuss in Section 8.3.2), they are volume or body forces (and so are drawn with their tip or tail at the body's center of mass).

These fictitious forces "deflect" the rock's sideways. The "deflect" is in quotes because fictitious forces, being fictitious, have no physical effect. Rather, the deflection is an artifact of the rotation (or acceleration) of the reference frame.

To see a fictitious force in action, we'll look at the cornering car (Section 7.3.2) anew, using a rotating, noninertial frame. The car, as a reminder, drives at a constant speed v on a circular road of radius R (Figure 8.2). For this reanalysis, it drives counterclockwise (because I like angles to increase) and experiences no air resistance (which would only obscure the upcoming fictitious force).

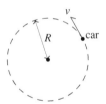

Figure 8.2 A car moving at constant speed in a circle. This car moves counterclockwise and, for simplicity, experiences no air resistance.

To make the rotating reference frame, stand at the center of the road and spin at the just right rate to hold the car directly ahead of you (Figure 8.3).

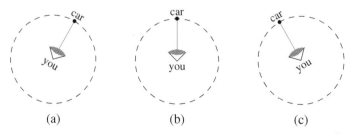

(a)   (b)   (c)

Figure 8.3 You at the center of the rotating reference frame. Standing at the center of the circle, you rotate yourself counterclockwise so that the car is always directly ahead of you. (a) An early moment. (b) A bit later. (c) Later still.

In the inertial frame, the car, being in uniform circular motion, has an acceleration of v^2/R inward (Figure 8.4a). This acceleration is caused by the two forces acting on the car: the gravitational force and the contact force. Their sum, the net force, is mv^2/R inward. Said another way, the contact force's normal portion balances the gravitational force, and its inward portion – static friction – provides the inward acceleration. Newton's second law is thereby satisfied.

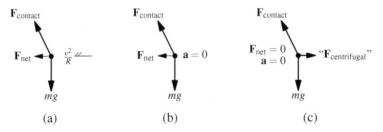

Figure 8.4 Freebody diagrams of the car in different reference frames (rear views, with the car driving into the paper and turning left to go around the circle). (a) In an inertial reference frame. (b) In your rotating reference frame. You stand at the origin, to the left of the car, and look to the right directly at the car – which, in your view, stays fixed. So, the car's acceleration, **a**, as seen in your frame, is zero, meaning that $\mathbf{F}_{net} \neq m\mathbf{a}$. (c) In the rotating reference frame of (b) but including a centrifugal force. With the correct $\mathbf{F}_{centrifugal}$ of (8.1), $\mathbf{F}_{net} = m\mathbf{a}$ (both are zero).

In the rotating frame, however, the second law is violated (Figure 8.4b). Although the freebody diagram has the same physical forces as in the inertial frame (physical forces are the same in all reference frames, inertial or otherwise), the acceleration **a** is now zero: In this convenient frame, the car doesn't move at all. Thus, the net force isn't $m\mathbf{a}$.

The solution is to invent a suitable fictitious force, used only in the rotating frame and acting on each body. This carefully constructed fictitious force is the centrifugal force (Figure 8.4c):

$$\mathbf{F}_{centrifugal} = m\omega^2 r \text{ outward,} \qquad (8.1)$$

where m is the body's mass, ω is the frame's angular rotation speed (as measured in an inertial frame), and r is the body's distance from the rotation axis.

What on earth is a reference frame's angular rotation speed (ω)?

It describes how rapidly the reference frame's orientation is changing (just as a frame's speed describes how rapidly the frame's position is changing). It is measured as angle (usually in radians) per time. For example, if the rotating frame requires 1 minute to make a complete rotation (an angle of 2π radians), then ω is 2π radians per minute or $\pi/30$ radians per second.

Does including the centrifugal force (8.1) rescue the second law?

To see whether this force offers what's needed, let's calculate it. The mass m is a given; and r, the car's distance from the rotation axis, is just R, the radius of the circular road. But ω is trickier. To calculate it, consider a duration t. In that time, the car travels along an arc of length vt (Figure 8.5). To keep the car straight ahead of you, your eyes must change their gaze direction by an angle $\theta = vt/R$ (an angle in radians is simply arc length divided by radius). Your angular rotation rate is the angle divided by the time. Thus,

$$\omega \equiv \frac{\theta}{t} = \frac{vt/R}{t} = \frac{v}{R}. \tag{8.2}$$

With those values, (8.1) gives

$$F_{\text{centrifugal}} = m \underbrace{\left(\frac{v}{R}\right)^2}_{\omega^2} R = \frac{mv^2}{R}. \tag{8.3}$$

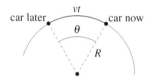

Figure 8.5 Calculating the rotation angle θ. In a time t, the car moves along the road a distance vt. To keep the car straight ahead of you, you rotate by $\theta = vt/R$.

Without the centrifugal force, the net force was mv^2/R inward – conveniently, just the opposite of the centrifugal force. Thus, the centrifugal force makes the net force zero. Because the car's acceleration, as measured in the rotating frame, is also zero, Newton's second law is now valid in the rotating frame!

To summarize, and in answer to the triangle question: If you apply Newton's second law in a rotating reference frame, you can – indeed, must – include a fictitious centrifugal force. This injunction complements my earlier injunction to avoid the centrifugal force completely (Section 1.5.1), an injunction still valid in an inertial frame.

Including fictitious forces in the net force creates a linguistic awkwardness. Because the net force now includes nonphysical forces, which cannot cause anything, it's sloppy to say that net force causes acceleration. I do it anyway, as if the centrifugal force had causal power, but remind myself of the sloppiness by placing quotation marks around $F_{\text{centrifugal}}$ on the freebody diagram (Figure 8.4c).

Although the centrifugal force's use is now legal, it remains a chainsaw, a powerful tool that can also slice off your leg. Its danger lies in the temptation to use it in an inertial reference frame. I am tempted myself because it seems to explain my feeling of being thrown outward against the car door as I drive around a turn. However, that feeling already has a physical force to explain it.

Without the car door (for simplicity, forget the seat belt and assume a frictionless seat), I would move in a straight line. The car and its door meanwhile are moving in a circle. Therefore, because of the inherently inward curve of a circle, the car door is moving and pushing me inward. I feel this contact interaction between me and the car door. On one side of the interaction, the car door pushes inward on me. On the other side, I push outward on the car door. But neither side is an outward force *on me*, which the centrifugal force would be.

Even so, my outward push on the car door tempts me into invoking the outward-pointing, centrifugal force. Such reasoning, however, mixes analyses from incompatible reference frames. The centrifugal force is legal only in a rotating reference frame, where it explains how I have zero acceleration (as seen in that frame) even though the actual net force on me is inward: The centrifugal force balances the contact force of the car door and makes the net force zero. However, when I import the centrifugal force into the inertial frame, I attempt to explain a phenomenon that needs no new force to explain it.

If, in a rotating reference frame, the centrifugal force is valid and needed, is it also valid to use the centripetal force?

Although the net force on the cornering car is centripetal (toward the center), no actual force is. Static friction does point toward the center, but it's not an actual force. It's only a portion of the actual contact force, and the contact force points upward and inward (for example, as in Figure 8.4a). Even the fictitious centrifugal force isn't centri*petal*. Thus, the reasons in Section 1.5.2 and Section 7.3.3 against using the centripetal force still apply. I avoid it in inertial and noninertial frames.

Is the centrifugal force the only fictitious force?

No! To see why, return to the ball dropped from the tower at the equator, as seen in the rotating-earth frame. Because of the rotation of the reference frame, the ball experiences a centrifugal force. But this force points outward – that is, upward along the tower. Thus, it cannot deflect the ball forward.

The centrifugal force, it turns out, is only one of several fictitious forces that arise because of the motion of the frame. These forces were invisible because the preceding reanalysis of the cornering car made three implicit assumptions. Relaxing each assumption reveals an additional fictitious force.

1. *Coriolis force.* The first assumption was that, in the rotating frame, the body is at rest – like the cornering car of Figure 8.4c. However, the falling rock of Figure 8.1 moves in the rotating frame, so it experiences a Coriolis force. This force points perpendicularly to the body's velocity v (as seen in the

noninertial frame) and to the frame's angular velocity ω (a vector that is itself perpendicular to the frame's plane of rotation). It has magnitude proportional to the frame's rotation rate ω, to the body's distance from the rotation axis, and to the body's speed v. This force deflects the falling rock forward.

2. *Euler force.* The second assumption was that the frame's rotation rate ω remains constant. A changing rotation rate creates the so-called Euler force. To see why this force is needed, return yourself to the center of the rotating frame used to reanalyze the cornering car (Figure 8.3) and, for a while, rotate at just the right rate to keep the car directly ahead of you. Then suddenly stop rotating (while the car continues its uniform circular motion). You, no longer rotating, see the car lurch forward. The car's acceleration, an artifact of the choice of reference frame, is produced by a correspondingly artificial or fictitious force – the Euler force. This force isn't as far-fetched as my extreme example makes it seem, for even the earth's rotation rate varies slightly. Thus, in our most natural reference frame, the rotating-earth frame, each body experiences a (tiny) Euler force.

3. *Frame-acceleration force.* The third assumption was that the frame's origin does not accelerate. Yet, any frame that accelerates, even if it doesn't rotate, is noninertial. Thus, in it, Newton's second law is invalid. For example, when you stand in the elevator accelerating upward in deep space (Figure 7.26), you have zero acceleration – as seen in the elevator frame. However, the net force on you isn't zero (it's mg upward). Thus, $\mathbf{F}_{\text{net}} \neq m\mathbf{a}$.

   We rescue the second law by adding a frame-acceleration force $-m\mathbf{a}_{\text{frame}}$ to each body, where $\mathbf{a}_{\text{frame}}$ is the frame's acceleration as measured in any inertial frame. Thus, to do physics in the accelerating-elevator frame, we add to you a fictitious force mg *downward*, making the net force on you zero.

So far, therefore, there are four fictitious forces: the centrifugal force, the Coriolis force, the Euler force, and the frame-acceleration force.

Are there further fictitious forces?

No! The list is finally complete at four. But avoid the subtle trap of considering fictitious forces as too similar to actual, physical forces. Although physical forces also fall into four categories (gravitational, electromagnetic, strong nuclear, and weak nuclear), those categories are the result of experiment. They could grow if new forces are discovered. As a child, I read many exciting newspaper articles about a fifth force. This force evaporated under the scrutiny of further measurements, but it was logically possible.

In contrast, the four fictitious forces result from mathematics – which isn't changing. Calculating the acceleration in the rotating frame always requires, in calculus terms, taking the derivative of the derivative of position. This double operation ends up producing four terms, each corresponding to one fictitious force. That mathematics will still hold if new physical forces are discovered or even if the universe ends. The four fictitious forces will always be four.

8.2 Torque and Rotation

If you've played tug-of-war (Section 5.6.1), your body knows that, if you stand upright, with static friction holding your feet in place, the rope will pull you head-over-heels forward. To avoid falling over, you lean backward. If you've skied (Problem 7.9), your body knows that, if you stand upright, you fall over backward. To avoid landing on your rear end, you lean forward.

However, falling over cannot be explained by Newton's laws. These laws, united in (7.44), describe how forces change the motion *of* a body's center of mass. Falling over, however, describes rotation *about* a body's center of mass.

In two dimensions, enough to illustrate the new ideas, *describing* rotation requires specifying two mathematical entities: (1) an axis of rotation – for example, the axis through a door's hinges or through a skier's center of mass – and (2) an angle of rotation about that axis.

Meanwhile, *explaining* rotation requires a new physical quantity: torque. Just as force measures pushing strength and causes acceleration (the second derivative of position), torque measures twisting strength and causes angular acceleration (the second derivative of angle).

Torque's relation to force is that a torque is produced by a force. Thus, the forces acting on a body affect not only (1) the motion of the body's center of mass but also (2) the body's rotation about its center of mass. To Newton's laws, which describe the first effect, we need to add two ingredients: how torque is calculated from force, and how it affects a body's rotation.

In general, torque is a pseudovector (as defined Section 5.7); however, in two dimensions, it can be considered a true scalar. Its sign says whether the force attempts to twist the body counterclockwise (positive torque) or clockwise (negative torque). Its magnitude is given by

$$|\,\text{torque}\,| \equiv |\,\text{force}\,| \times \text{lever-arm length}, \qquad (8.4)$$

where the *lever-arm length* is the distance between the axis of rotation and the force's line – not its point – of application. (The distance between two lines is their perpendicular distance.) For example, for a partly open dishwasher door pulled horizontally (Figure 8.6), the lever arm is the perpendicular distance from the axis (the hinges) to the dotted line – not the distance along the door.

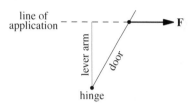

Figure 8.6 A dishwasher door being pulled open (side view).

With the definition of torque (8.4), its effect on rotation can be formalized:

$$\frac{1}{\text{moment of inertia}} \times \text{net external torque} \rightarrow \text{angular acceleration}, \quad (8.5)$$

where the *moment of inertia* is the rotational analog of mass, and *angular acceleration* is the rotational analog of acceleration (it's the rate of change of *angular* velocity). Moment of inertia depends not only on a body's mass but also on how dispersed its mass is from the axis of rotation.

The causal equation (8.5) is the rotational analog of Newton's second-and-a-half law (7.44). For force, only external influences mattered because of Newton's third law (Section 7.2). For torque, only external influences matter because of a stronger form of the third law than I stated on p. 1 and we've been working with so far. In the usual, weaker form, an interaction consists simply of two forces that are equal and opposite, so they cancel out in the net-force vector sum (which includes the forces on both bodies in the interaction). However, their respective torques might not cancel – which is what the strong form of Newton's third law forbids. It says that the two forces not only point in opposite directions but also have a common line of action. Thus, their lever arms are equal, making their respective *torques* equal and opposite – so they cancel out in the net-torque sum. Thus, in (8.5), we need to compute only external torques (torques produced by external forces).

Just as Newton's second law is easily misunderstood as saying that force produces velocity rather than acceleration, torque is easily misunderstood as producing angular velocity. Instead, just as force changes velocity, torque changes angular velocity. Thus, just as constant velocity requires no force for its explanation, constant angular velocity requires no torque for its explanation. Constant velocity implies zero net force; constant angular velocity implies zero net torque.

As an example, launch a stick as far as you can (Figure 8.7), perhaps for a dog to play fetch. The stick's center of mass follows a curved trajectory (a parabola) due to the only external force, the gravitational force (ignoring drag).

Figure 8.7 A thrown stick. Its center of mass moves in a parabola, as any point particle would (without drag). Meanwhile, it spins around its center of mass.

▶ *How does this force affect the stick's rotation about its center of mass?*

Because this force acts at the stick's center of mass, which is the rotation axis, its lever arm is zero. From the torque's definition (8.4), zero lever arm means zero torque. From torque's effect on rotation (8.5), therefore, the stick's angular velocity remains constant at its launch value.

To launch a stick, I grip the bottom end, move my arm and hand rapidly in an arc, and, as I let go, flick my wrist and the stick forward. The stick gets launched rotating about its center of mass with topspin. As it moves along its trajectory, it keeps the same rotation sense and rate. (Drag would produce a counterclockwise torque and reduce the rate slowly.)

A famous cinematic depiction is an early scene in Stanley Kubrick's film *2001: A Space Odyssey* (1969), where the inspired ape, our presumed ancestor, launches a large bone high into the air. The bone moves, spins, and suddenly turns into a spaceship – a culmination of the development of many tools, not least Newton's laws.

After the preceding introduction to torque and rotation, we can use these ideas to calculate how to stand while skiing. For simplicity, make the hill frictionless. Furthermore, I volunteer myself as the skier because I skied terribly until I understood the following physics (now I ski merely badly). Imagine that I have just started moving downhill. My speed is low, so air resistance, proportional to speed squared (Section 1.3.3), is small enough to neglect. Thus, as when you slid down a frictionless hill (Section 7.1.4), only two forces act on me, the normal (contact) force and the gravitational force.

▶ *How do these forces affect me if I stand upright?*

In Section 7.1.4, we found their effect on my center of mass: It accelerates downhill with acceleration magnitude $g \sin \theta$ (where θ is the hill's inclination). The resulting thrills are the reason to ski.

But do the thrills last: Do I remain standing or fall over? That question turns on, so to speak, whether I start turning, or rotating, about my center of mass – which, from (8.5), depends on the net external torque acting on me.

To calculate this torque, we calculate, for each external force acting on me, its corresponding torque using (8.4). These calculations require the lever-arm length, which requires specifying the rotation axis: through my center of mass and perpendicular to the page. Rotation about this axis spells trouble: falling over. (For slightly complicated reasons, the rotation axis should either be a fixed axis, which here isn't convenient; or, if not fixed, go through the center of mass.)

For the gravitational force, which acts at my center of mass, the lever-arm length is zero, so this force contributes no torque. Meanwhile, the torque that the normal force contributes depends on how I stand.

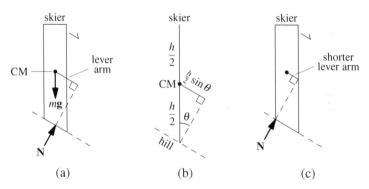

(a) (b) (c)

Figure 8.8 Skiing while standing upright. (a) The freebody diagram. The normal force tries to rotate me counterclockwise around my center of mass. The gravitational force, meanwhile, has zero lever arm, so it cannot save me from the normal force's torque. I fall on my rear end. (b) Calculating the lever-arm length in (8.6). The length is the altitude of a right triangle with hypotenuse $h/2$. (c) Shifting my stance uphill. This shift shortens the lever arm, but not to zero. I still fall backward.

When I stand upright (Figure 8.8a), my center of mass lies behind the normal force's line of action. Because the lever arm isn't zero, the normal force provides a nonzero torque, one that tries to rotate me ever-faster counterclockwise. Thus, in answer to the triangle question, I fall backward onto my rear end.

Roughly how long does that process take?

From (8.5), the torque produces an angular acceleration proportional to the torque and inversely proportional to my moment of inertia. Thus, we'll first estimate (1) the torque on me and (2) my moment of inertia, in order to estimate (3) my angular acceleration.

1. *Torque.* For a skier with height h going down a hill of slope θ,

$$\text{lever-arm length} = \frac{h}{2}\sin\theta :\qquad (8.6)$$

the factor of one-half because the center of mass is halfway up the skier and the factor of $\sin\theta$ because the lever arm is the altitude of the right triangle with opening angle θ (Figure 8.8b). For me, h is roughly 2 meters (less roughly if I had eaten a more as a child). An easier intermediate ski slope has $\theta \approx 15°$ (a 25-percent slope, where the percentage measures $\tan \theta$). With those values, the lever arm is roughly 0.25 meters long. The normal force on me is $mg \cos \theta$ or, at $m = 60$ kilograms, roughly 600 newtons. Thus, from (8.4), the normal force contributes a torque of 150 newton meters counterclockwise.

2. *Moment of inertia.* For a thin rod (a reasonable approximation to a person) rotating about its center of mass, the moment of inertia I turns out to be

$$I = \frac{1}{12}mh^2, \tag{8.7}$$

where m is the rod's mass, and h is its length. For me,

$$I \approx \frac{1}{12} \times 60\,\text{kg} \times (2\,\text{m})^2 = 20\,\text{kg m}^2. \tag{8.8}$$

3. *Angular acceleration.* Thus, my angular acceleration, symbolized by α, is

$$\alpha = \frac{\text{torque}}{I} \approx \frac{150\,\overbrace{\text{N}}^{\text{kg m s}^{-2}}\,\text{m}}{20\,\text{kg m}^2} \approx 7\,\text{rad s}^{-2}. \tag{8.9}$$

(The "rad" is the abbreviation for radians, of which 2π make a full circle.)

Now we can estimate how long I experience this angular acceleration before hitting the ground. In this situation of constant (angular) acceleration, my rotation angle increases quadratically with time, like the distance that a stone starting from rest falls in free gravitational motion. Analogous to (7.2) for the stone,

$$\theta = \frac{1}{2}\alpha t^2. \tag{8.10}$$

Falling over means rotating by an angle of, say, 1 radian (roughly 60 degrees), so it requires just 0.5 seconds:

$$\frac{1}{2} \times \underbrace{7\,\text{rad s}^{-2}}_{\alpha} \times \underbrace{(0.5\,\text{s})^2}_{t^2} \approx 1\,\text{rad}. \tag{8.11}$$

In my experience of skiing and often falling backward, this estimated time feels about right. It's long enough to be perceptible: I was always aware that I was falling backward. It's also short enough to make falling over unavoidable: In the available time, I could never prevent the coming thump.

▶ *How can falling over be prevented?*

Preventing it means eliminating the angular acceleration, which means eliminating its cause: the normal force's torque. This torque's magnitude is the product of the normal force's magnitude N and the lever-arm length. Although N is difficult to reduce, I can shorten its lever arm in two ways.

First, using the approach of many beginning skiers (including me), I can shift my stance uphill to my heels – that is, I try to rotate myself clockwise by pushing against the ground with my heels. The normal force then acts at my lower left corner (Figure 8.8c). This shift does shorten the normal force's lever arm. However, unless I am extremely squat (sledding!), it cannot shorten the lever arm to zero. Thus, I still fall backward.

Second, I can bring my center of mass forward, toward the normal force's line of action. When my center of mass lies on this line, the lever-arm length is zero. More simply described: I lean forward until I stand perpendicularly to the hill (Figure 8.9). Now I can ski without falling backward.

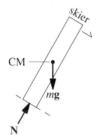

Figure 8.9 My freebody diagram standing perpendicular to the hill. Now the normal force has zero lever arm, so it doesn't topple me.

The preceding knowledge of lever arms has been incorporated into earlier freebody diagrams – for example, for bicycling on level ground with zero acceleration (Figure 5.6c). The composite body of you and the bicycle maintains a fixed rotation rate (zero) about its center of mass. Thus, the net external torque on it must be zero. And it's zero if each force's line of action passes through the center of mass. For this reason, I drew the contact force of the road acting slightly behind the center of the bottom surface. (Its location raises an enjoyable puzzle, which I leave to you, extending the discussion of Section 7.1.6: How does a passive force figure out its point of application?)

8.3 Going beyond Newton's Laws

Newtons laws of motion, despite the effort needed for their mastery – I'm learning after 30 years of study – aren't absolute truths. In this final section, I mention their limitations and the physical ideas developed to address them.

8.3.1 Changing Mass

The first limitation, easier to fix, is that the second-and-a-half law (7.44) assumes that the body's mass M is constant. Although a point particle cannot change its mass (or else it would have internal structure and not be a point particle), a composite body might lose or gain mass.

An example is a rocket in deep space. There, the composite body of the rocket and its fuel is isolated, meaning that all interactions in which its constituents participate are within the composite body. Thus, the net external force on the composite body is zero, as is its acceleration.

Yet, the rocket accelerates anyway. The reason is that the exploding, burning, and expanding fuel exerts a forward force on the rocket (just as, by the third law, the rocket exerts a backward force on the fuel). This force, the only external force on the rocket, causes the rocket to accelerate.

But, in applying the second-and-a-half law (7.44) to the rocket, to what body does the mass M refer? Is it the rocket itself, whose mass is constant? Is it all the fuel, whose mass is different from the rocket's but also constant? Is it the rocket with the fuel remaining in the rocket, whose mass is changing?

This question is resolved by analyzing one tiny time step at a time. At each step, you divide the universe into two bodies: (1) the tiny bit of fuel to be ejected during this time step and (2) the rocket and its remaining fuel (excluding the tiny portion to be ejected). During the time step, each body's mass is fixed; thus, Newton's second-and-a-half law is valid. Knowing how fast the fuel is ejected relative to the rocket (the fuel's exit speed), you use the second-and-a-half and third laws to determine the interaction strength between, and thus the forces acting on, the two bodies. The eventual result is a differential equation for the rocket's speed as a function of how much fuel remains.

In a simpler resolution, Newton's laws are reformulated using a new physical quantity: momentum. A vector quantity, it's symbolized as $\mathbf{P}$ and described as "quantity of motion" – the physical quantity that I introduced in Section 7.1.2 to distract our intuition from thinking too much about velocity. For a body with mass m and velocity $\mathbf{v}$,

$$\mathbf{P} \equiv m\mathbf{v}. \tag{8.12}$$

Newton's second law is the special, constant-mass case of the general statement that force is the rate of change of momentum: that $\mathbf{F} = d(m\mathbf{v})/dt$. When m is constant, $\mathbf{F}$ simplifies to $m\,d\mathbf{v}/dt$, which is our old friend $m\mathbf{a}$ of the second law.

And the principle of (local) conservation of momentum – that momentum is neither created nor destroyed but only flows from one body to another – means that, in an interaction between two bodies, the rate at which momentum leaves one body equals the rate at which momentum arrives at the other body. Thus, the corresponding forces are equal and opposite – giving Newton's third law.

For the rocket–fuel system, each tiny bit of escaping fuel carries away momentum – an amount that is easy to compute from the mass of the piece and its exit speed. This momentum comes from the rocket (and its remaining fuel). Equating the momentum changes then quickly gives the same differential equation for the speed of the rocket as the longer, trickier analysis with forces gives.

Thus, for the rocket, momentum is a convenient tool, but you can use forces if you think carefully. In more advanced physics, such as quantum mechanics and relativity, however, momentum plays the fundamental role; forces are computed, if at all, from flows of momentum.

8.3.2 Speed-Of-Light Limit

The second, more fundamental limitation of Newton's laws lies in the instantaneous nature of the third law: It requires that the forces making up an interaction be equal and opposite always. The problem hidden therein isn't apparent with short-range interactions but becomes apparent with long-range interactions.

The electromagnetic interaction, though short range in the examples in this book, is also long range – think of light from the sun. To turn this interaction into a problem for Newton's laws, arrange three charges as follows (Figure 8.10a). Place a positive charge on the sun. Place a unit negative charge at the earth's south pole and a unit positive charge at the earth's north pole (pretending that the earth's rotation axis isn't tilted).

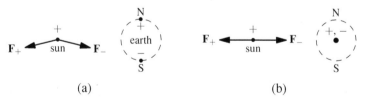

(a)   (b)

Figure 8.10 The forces on a positive charge on the sun, due to positive and negative charges on the earth. (a) The earth charges at the poles and separated vertically. The forces on the sun charge don't balance. (b) The earth charges brought together on the equator. Now the forces on the sun charge balance.

With this arrangement, the sun charge participates in two interactions: an attractive Coulomb's-law interaction with the negative, south-pole charge and a repulsive Coulomb's-law interaction with the positive, north-pole charge. The two resulting forces have the same magnitude: As given in (1.30), the magnitude depends only on the distance between the charges and on the charge magnitudes (as would the gravitational force though with charge replaced by mass). The forces point in mostly opposite directions, but with the slight difference that both point slightly downward. The resulting net force on the sun charge is downward and, most importantly, nonzero.

Now quickly bring the north- and south-pole charges together at the center of the earth (Figure 8.10b) – in, say, 1 minute (each charge then moves at 100 kilometers per second). Once the positive and negative charges are exactly on top of each other, the two forces on the sun charge point in exactly opposite directions. Thus, the net force on the sun charge is now zero.

Information about the earth charges' relative position has thereby been sent from the earth to the sun, a distance of 150 million kilometers, in 1 minute. This signal traveled at roughly eight times the speed of light! By placing the movable charges closer and raising the speed at which they are brought together, that signal-propagation speed could be increased arbitrarily. Or so it seems from this analysis based only on Coulomb's law.

However, a complete analysis requires Maxwell's equations [4], only one of which is Coulomb's law. In the complete analysis, each earth charge's changing location is signaled by a wave that travels from the charge to the sun charge. This wave, which carries momentum, travels through a new physical entity, the electromagnetic field, whose behavior is described by Maxwell's equations. Their consequence is that a changing electric field creates nearby a changing magnetic field, which creates nearby a changing electric field. The fields chase their tails all the way across the 150 million kilometers from the earth to the sun. These changes travel at roughly 3×10^5 kilometers per second – the speed of light c. Thus, when analyzed carefully, the electromagnetic interaction and the resulting forces (or momentum transfers) cannot send signals arbitrarily fast.

As so often happens, resolving one problem creates another – thus is paradox the engine of progress. Here, the new problem is the inconsistency between a speed limit for electromagnetic signals versus no speed limit for bodies themselves. For if you keep pushing a body in one direction, it keeps accelerating and eventually moves faster than the speed of light. Then you could use the body itself to send signals faster than the speed of light (just tape a message to the body) and circumvent the electromagnetic speed limit. Thus, electrodynamics, represented by Maxwell's equations, and Newtonian dynamics, represented by Newton's laws (in particular, the second law), are inconsistent.

This problem was resolved by Einstein with the theory of special relativity (my favorite treatment of it is by Edwin Taylor and John Wheeler [22]). A short summary of the resolution is that a body's momentum, formerly defined in

(8.12) as $m\mathbf{v}$, becomes

$$\mathbf{P} \equiv m\mathbf{v}\gamma, \tag{8.13}$$

where the new factor γ is

$$\gamma \equiv \frac{1}{\sqrt{1 - v^2/c^2}}. \tag{8.14}$$

Force remains the rate of change of momentum. But now, thanks to the γ factor, which goes to infinity as v approaches c, P can approach infinity without the speed v increasing beyond the speed of light c. Thus, a steady force can increase P forever, yet v never exceeds c.

Fortunately for Newtonian mechanics, implicitly based on the $m\mathbf{v}$ definition of momentum (embedded in Newton's second law), the new γ factor matters only for bodies moving at speeds comparable to the speed of light (or for slowly moving bodies where extreme precision is needed, such as GPS satellites). Almost every object in everyday life, on earth and even in the solar system, moves slowly compared to light, making v^2/c^2 almost zero and γ almost exactly 1. Thus, Newton's laws remain a useful approximation.

With the γ factor of special relativity, the speed of light is baked not only into electrodynamics but also into mechanics itself, making mechanics and electrodynamics consistent. However, this resolution, as Einstein recognized soon after developing special relativity, created a problem for the other long-range interaction, gravitation.

To see the problem, consider the gravitational force on the sun due to the earth as the earth orbits the sun (Figure 8.11). This force, according to the law of universal gravitation (1.2), always points directly from the sun to the earth. Assume that, right now, it does.

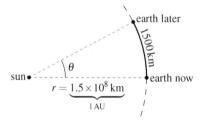

Figure 8.11 The earth traversing a tiny arc of its orbit (50 seconds worth).

However, look what happens during the next 50 seconds. The earth orbits at 30 kilometers per second; in those 50 seconds, it moves 1500 kilometers along an arc of its solar orbit. That distance corresponds to an angle of 10^{-5} radians:

$$\theta \text{ (in radians)} \equiv \frac{\text{arc length}}{\text{radius } r} = \frac{1500 \, \text{km}}{1.5 \times 10^8 \, \text{km}} = 10^{-5}. \tag{8.15}$$

The earth's gravitational force on the sun must also rotate by the same (small) angle. Thus, in these 50 seconds, the earth has sent a signal to the sun about its new position. This signal traversed 500 light seconds (the earth–sun distance), so it has traveled at 10 times the speed of light. Thus, allowing instantaneous gravity opens a new loophole in the universal speed limit of c.

The analogous problem for charges, where Coulomb's law is an inverse-square law like gravity, was fixed with Maxwell's equations and the idea of an electromagnetic field. But what are the field equations of gravity, analogous to Maxwell's equations, that could close this loophole?

Einstein developed them using a surprising idea: He eliminated gravity. He noticed that one can always choose a noninertial reference frame where the fictitious frame-acceleration force $-m\mathbf{a}_{frame}$ on each body (fictitious force 3 on p. 179) exactly balances the gravitational force, which is also proportional to m. Choosing that frame seems complicated. However, it just means choosing a frame whose acceleration $\mathbf{a}_{frame}$ equals the gravitational field $\mathbf{g}$. Then

$$-m\mathbf{a}_{frame} + m\mathbf{a} = 0. \qquad (8.16)$$

And it has a simple physical implementation: Go into free gravitational motion (free fall). In a free-fall frame, gravity gets balanced away by the frame-acceleration force. With gravity's disappearance goes also any problem of its infinite propagation speed!

As you might suspect by now, that resolution created a new problem. These noninertial frames, in order to fall freely, have to be tiny (or, in physics jargon, local) because these tiny, freely falling frames near any gravitational source have different accelerations (Figure 8.12). Thus, these frames cannot be stitched together into a larger accelerating frame (such as the noninertial frames of Section 8.1, which implicitly provide a vantage point for the whole universe).

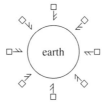

Figure 8.12 Local inertial frames. Although the tiny freely falling elevators are inertial frames, they cannot be stitched together into a global inertial frame.

Restricted to only tiny noninertial frames, how do observers in separate frames compare measurements or make a global picture of the universe? For example, how can they even state that a rock is moving in a straight line at constant speed, if each observer knows only that it moves at constant velocity in her tiny frame?

Stitching together these local measurements required Einstein's next idea: that space (and time) are curved. The long-sought-for field equations of gravitation describe how matter curves space (and time) and, along with the so-called geodesic principle, how signals propagate and bodies move – all while obeying the speed-of-light limit. This amazing resolution is Einstein's theory of general relativity. (My favorite treatment is again by Edwin Taylor and John Wheeler [23].) It makes many predictions that have been confirmed over the past 100 years [26, 27] – including the existence of gravitational waves, first observed in 2015 [1]. At the same time, it creates a new problem: to combine this theory of gravity – with its omnipresent, ever-changing gravitational field that also provides the stage on which all physics happens – with quantum mechanics, the theory of the microscopic world. That problem awaits a solution. Perhaps you will be inspired to work on it.

Although it may seem sad to end our time together with the limits to Newton's laws, the truth is the opposite. By exploring the limits in Newton's laws, humanity has been led to new theories and new models of the universe. The limitations of one generation's knowledge kindle the advances of the next.

8.4 Bon Voyage!

Galileo's law of inertia, which became Newton's first law, and Newton's second and third laws of motion formed the spark that, in the 17th century, set the scientific revolution alight. They overturned millennia of revered authority, not least that of the polymath genius Aristotle and of other ancient Greeks. Yet, cultural change is powerless alone. For culture, including scientific culture, is carried in individuals, who, working together, comprehend and change our world. Thus, the scientific revolution must be made anew, in each generation and in each of us. By mastering Newton's laws, you are making that revolution within you. May it be one step of many on our way to understanding the universe, a journey that has no end.

8.5 Problems

8.1 Return to the cornering car of Section 7.3.2 in uniform circular motion. Here, ignore air resistance, which means that the static-friction force **f** points directly inward.

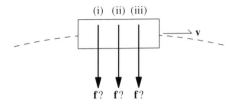

Figure 8.13 Three possible locations of the frictional force on the cornering car: (i) before, (ii) through, or (iii) ahead of the center of mass.

    a. Where does f's line of action lie? Does it pass behind the car's center of mass [choice (i) in Figure 8.13], through it [choice (ii)], or ahead of it [choice (iii)]?

    b. When the car, still in circular motion, accelerates (in the everyday sense of speeding up), where does f's line of action lie?

    c. When the car decelerates, where does f's line of action lie?

In choosing the correct line of action, you can ignore the gravitational and normal forces. Being vertical forces, they don't affect the car's rotation in the horizontal plane (they turn the car neither left nor right).

8.2 In introducing skiing downhill on the wedge of Section 7.4.3, I mentioned that you must crouch or squat in order to avoid falling over backward. As a simple model of this situation, imagine yourself as a uniform block with width w and height h. As a function of the inclination, θ, what is the maximum aspect ratio h/w that you can have without falling backward? You can check your result approximately by comparing it to the m block in Figure 7.23, which has been drawn with that maximum aspect ratio.

8.3 Reanalyze the forces on and the acceleration of the maximum-aspect-ratio block of Problem 8.2 but working as follows in the noninertial reference frame of the wedge or block.

    a. In this frame, what is the block's acceleration? (Don't overthink it!)

    b. Which fictitious force, of the four in Section 8.1, must be included on the freebody diagram? Find its magnitude and direction, and include it on the block's freebody diagram (acting at the center of mass).

    c. Redraw the diagram combining the fictitious force and the gravitational force into a single "effective gravitational force" $\mathbf{F}_{g\,\text{effective}}$. How are $\mathbf{F}_{g\,\text{effective}}$ and the contact force $\mathbf{F}_{\text{scale}}$ related (in direction, magnitude, and line of action)?

    d. In this frame, which direction is "up" (the direction opposite to the effective gravitational force)? To avoid falling backward when skiing (analyzed in Section 8.2), you should stand pointing in this direction.

References

[1] B. P. Abbott, R. Abbott, T. D. Abbott, et al. Observation of gravitational waves from a binary black hole merger. *Physical Review Letters*, 116:061102, 2016.

[2] Florian Cajori, editor. *Sir Isaac Newton's Mathematical Principles of Natural Philosophy and His System of the World*. University of California Press, Berkeley, CA, 1966.

[3] Stillman Drake. *History of Free Fall: Aristotle to Galileo with an Epilogue on π in the Sky*. Wall & Thompson, Toronto, 1989.

[4] Daniel Fleisch. *A Student's Guide to Maxwell's Equations*. Cambridge University Press, Cambridge, UK, 2008.

[5] Simon Gindikin. *Tales of Mathematicians and Physicists*. Springer Verlag, New York, 2007.

[6] Robert Goddard. How my speed rocket can propel itself in a vacuum. *Popular Science Monthly*, page 38, August 1924.

[7] Cornellis Hellingman. Newton's third law revisited. *Physics Education*, 27(2):112–115, 1992.

[8] Paul G. Hewitt. *Conceptual Physics*. Pearson, Boston, 12th edition, 2015.

[9] E. T. Jaynes. Probability theory as logic. In Paul F. Fougere, editor, *Maximum Entropy and Bayesian Methods*, pages 1–16. Kluwer Academic Publishers, Dordrecht, NL, 1990.

[10] E. T. Jaynes. A backward look into the future. In W. T. Grandy Jr. and P. W. Milonni, editors, *Physics and Probability: Essays in Honor of Edwin T. Jaynes*, pages 261–275. Cambridge University Press, Cambridge, UK, 1993.

[11] Bob Kibble. Understanding forces: What's the problem? *Physics Education*, 41(3):228–231, 2006.

[12] Jill Larkin, John McDermott, Dorothea P. Simon, et al. Expert and novice performance in solving physics problems. *Science*, 208(4450):1335–1342, 1980.

[13] Sanjoy Mahajan. *Street-Fighting Mathematics: The Art of Educated Guessing and Opportunistic Problem Solving*. MIT Press, Cambridge, MA, 2010.

[14] Sanjoy Mahajan. *The Art of Insight in Science and Engineering: Mastering Complexity*. MIT Press, Cambridge, MA, 2014.

[15] Mars Climate Orbiter Mishap Investigation Board. Phase I report. Technical Report, NASA, 1999.

[16] Judea Pearl and Dana MacKenzie. *The Book of Why: The New Science of Cause and Effect*. Basic Books, New York, 2018.

194

[17] George Pólya. Let us teach guessing: A demonstration with George Pólya [videorecording]. Mathematical Association of America, Washington, DC, 1966.

[18] Frederick Reif. Millikan Lecture 1994: Understanding and teaching important scientific thought processes. *American Journal of Physics*, 63(1):17–32, 1995.

[19] Frederick Reif and Sue Allen. Cognition for interpreting scientific concepts: A study of acceleration. *Cognition and Instruction*, 9(1):1–44, 1992.

[20] Alexander Renkl. Learning from worked examples: How to prepare students for meaningful problem solving. In Victor A. Benassi, Catherine E. Overson, and Christopher M. Hakala, editors, *Applying Science of Learning in Education: Infusing Psychological Science into Curriculum*, pages 118–130. Society for the Teaching of Psychology, Washington, DC, 2014.

[21] Angus Stevenson and Christine A. Lindberg, editors. *New Oxford American Dictionary*. Oxford University Press, New York, 3rd edition, 2010.

[22] Edwin F. Taylor and John Archibald Wheeler. *Spacetime Physics*. W. H. Freeman, New York, 2nd edition, 1992.

[23] Edwin F. Taylor and John Archibald Wheeler. *Exploring Black Holes: Introduction to General Relativity*. Addison Wesley Longman, San Francisco, 2000.

[24] Colin Terry and George Jones. Alternative frameworks: Newton's third law and conceptual change. *European Journal of Science Education*, 8(3):291–298, 1986.

[25] J. W. Warren. *Understanding Force: An Account of Some Aspects of Teaching the Idea of Force in School, College, and University Courses in Engineering, Mathematics and Science*. John Murray, London, 1979.

[26] Clifford M. Will. *Theory and Experiment in Gravitational Physics*. Cambridge University Press, Cambridge, UK, revised edition, 1993.

[27] Clifford M. Will. *Was Einstein Right? Putting General Relativity to the Test*. Basic Books, New York, 2nd edition, 1993.

Index

An italic page number refers to a Problem on that page.

Printed in the United States
By Bookmasters